Halford Mackinder

Halford J. Mackinder, oil portrait.
Courtesy of the University of Reading.

Brian W. Blouet

HALFORD
MACKINDER
A Biography

Texas
A&M University
Press
COLLEGE STATION

Library of Congress Cataloging-in-Publication Data

Blouet, Brian W., 1936–
   Halford Mackinder : a biography.

   Bibliography: p.
   1. Mackinder, Halford, Sir, 1861–1947. 2. Geographers
—Great Britain—Biography. I. Title.
G69.M23B58 1987   910'.92'4 [B]   86-23013
ISBN 0-89096-292-8

The paper used in this book meets the minimum requirements
of the American National Standard for Permanence
of Paper for Printed Library Materials, Z39, 48-1984.
Binding materials have been chosen for durability.

Contents

List of Illustrations

Preface

On January 31, 1943, Sir Halford Mackinder wrote to the director of the London School of Economics, A. M. Carr-Saunders, declining an invitation to answer a questionnaire on the early years of LSE. Mackinder added that he was working on a book that would cover the major ventures of his life and he hoped to have it published in the coming year. In fact, he had been organizing materials and writing passages for an autobiography over the last twenty years of his life. But this man who possessed extraordinary foresight had difficulty looking back. The autobiography was never finished, although parts of Mackinder's typescript exist.

Work on this biography of Mackinder started almost by accident in the summer of 1974 at the School of Geography, Oxford. I was beginning a research leave from the University of Nebraska and working at the School at the suggestion of Professor Jean Gottmann. I casually mentioned an interest in Mackinder at coffee one morning, and Harold Jefferies, administrator to the School, said there was a large box of materials in his office. Would I like to see them? A black tin trunk was taken into the library and with the help of the School's librarian, Elspeth Buxton, I sorted the papers into general categories and listed them. Many of the papers were family letters, written after Mackinder's death, and they gave leads as to where additional material could be found. The overall result was to gather together fresh information about Mackinder's life and interests.

Several other visits to the School followed, and in 1980–81 I was again given research leave by the University of Nebraska and spent the year in Oxford drafting this book. At the School of Geography, I am indebted to David Browning, Elspeth Buxton, Paul Coones, Jean Gottmann, Andrew Goudie, John House, William Mills, Ceri Peach, Gordon Smith, and Marjorie Sweeting for innumerable discussions in which ideas about Mackinder were opened out. Ken Warren was particularly helpful on Mackinder's business interests and put me onto the material at Vickers Limited. Peter Masters photographed several of the illustrations.

Jean Gottmann, John Patten, and Andrew Goudie took me along to Hertford College and made me a part of Senior Common Room life during my visits. All universities have mechanisms to allow scholars from different fields to exchange ideas, but the college Senior Common Room is very effective and certainly the most convivial.

Away from Oxford, many others have helped and encouraged. Alice Garnett and the late Charles Fisher, who gave me my first academic appointment at the University of Sheffield, both took an interest in the work. Alice provided recollections of Mackinder, and Charles shared ideas on Mackinder's view of world affairs. Leonard Cantor, Schofield Professor of Education, University of Technology, Loughborough, was always encouraging and generously shared with me materials in his possession. David Hooson, Berkeley, was helpfully critical and pushed me along with the work. Walter Freeman, busily retired in Abingdon and a frequent visitor to Oxford, was marvelous. He may not agree with me on everything concerning Mackinder, but with great generosity he made available an important file of material which he had collected over thirty years. The materials included letters from Fleure, Fairgrieve, and others containing recollections of Mackinder.

I have been greatly aided by many kindly librarians, archivists, and officers of institutions where materials relating to Sir Halford Mackinder were found or might have been found. At the Bodleian Library, David Vaisey made available materials from the university archives. Dr. James Williams and Michael Bott, archivists at the University of Reading, were always helpful. At the London School of Economics, Mr. Bovey and Miss Sanders produced documents from the library and the archives. Mr. Mellor, of the Foreign and Commonwealth Office, unearthed the references to the Colonial Office Visual

Instruction Committee, and Donald Simpson, librarian, Royal Commonwealth Society, introduced me to the A. Hugh Fisher materials commissioned by the Visual Instruction Committee. Christine Kelly of the Royal Geographical Society has a very great knowledge of RGS documents and produced very many unexpected and valuable papers from the Society's holdings. A. J. Trump, Rewley House, guided me through the Oxford Extension archives, and Mr. Wing and Mrs. Bradshaw of the Library, Christ Church, extracted material from the Governing Body minutes. J. S. English of the Lincolnshire Libraries provided me with transcripts of the 1861 and 1871 censuses relating to Gainsborough and took the time to show me around Mackinder's hometown. The Staffordshire Record Office was particularly helpful in finding directory references to the Hewit family of Lichfield, and the staff of Epsom College were very pleasant in making records available for study.

David Hooson, Jean Gottmann, David Wishart, and Geoffrey Martin read the typescript and made many valuable suggestions, which greatly improved the text. Geoffrey Martin is another scholar who has given me encouragement over the years. The University of Nebraska was generous in its provision of research leaves and funds. Among the Nebraska students, Stephen Riley and David Benorden turned up published materials which were of great value, and David shared his knowledge of German geopolitics with me. The maps and the diagram of Mackinder's career which appear in this volume were prepared by the Cartographic Services Unit, Department of Geography, Texas A&M University, under the direction of Christopher Mueller-Wille.

Finally, I acknowledge the help of my wife, Dr. Olwyn Blouet, a historian of the British Empire-Commonwealth, for her editorial work and encouragement throughout the project.

List of Abbreviations

BLPES   British Library of Political and Economic Science,
             London School of Economics

Bodleian   Bodleian Library, Oxford

HFA   Hamilton Fish Armstrong Papers, Princeton
             University

HP   Hewins Papers, University of Sheffield

LAO   Lincolnshire Archive Office

LSE   London School of Economics, Records in
             Connaught House

M.P. Auto.   Mackinder Papers, Autobiographical Fragments,
             School of Geography, Oxford

PRO   Public Record Office, London

REC   Records of Epsom College

RGS   Royal Geographical Society

RHO   Rewley House, Oxford

RUA   Reading University Archives

SNL   Scottish National Library, Edinburgh

UCA   University of Chicago Archives

Halford Mackinder

1

Lincoln and Lichfield

This is the story of an intelligent and imaginative man—
Halford Mackinder—who was born in a small Lincolnshire
town and rose, by virtue of talent and hard work, to be one
of the leading educators of his day. Mackinder established a
new discipline, geography, in English universities, laid the foundations for a university at Reading, made the first ascent of
Mount Kenya, analyzed Britain's economic problems and predicted the decline of the country in world affairs. In a famous
statement entitled the "Geographical Pivot of History" (1904)
he announced the probable emergence of a great landpower
dominating Eurasia, and later, in his book *Democratic Ideals
and Reality* (1919), he warned of the rise of totalitarian regimes.

Halford Mackinder was born on February 15, 1861, at Gainsborough, a small port, and market town, on the banks of the
River Trent. The name Mackinder is not uncommon in Lincolnshire; in fact, there is a surprisingly large number of names
of Scottish origin in the county. From the fifteenth century
on, Scottish fishermen and traders were working the east coast
of England, selling fish and buying malt and barley. As opportunities for settlement opened up on the margins of marshes
and fens, Scotsmen became farmers or smallholders. By the
seventeenth century many families of Mackinders were recorded as paying hearth tax in Lincolnshire.[1]

Halford's father believed that his branch of the family had
come to Lincolnshire as fugitives from the 1745 rebellion in
Scotland. Whatever their motive, they arrived with sufficient
capital to go into farming in a substantial way, and they rented

1. Transcripts of the Hearth Tax compiled by Canon Foster, Lincolnshire Archive Office (LAO).

extensive tracts of newly enclosed land. John Mackinder, one of the reputed fugitives, appears in the mid-eighteenth century married to Eleanor Draper of Rowston. The couple had six children, the last of whom was Halford's great-grandfather Draper Mackinder (1751–99). The first Draper Mackinder became a substantial farmer with land around the villages of Rowston and Timberland. A lease of 1791 shows the first Draper Mackinder farming 330 acres of arable land and pasture and paying an annual rent of £123.[2] By this time farmer Mackinder had married Elizabeth Dawson of Lincoln. Their first child, born in 1782, was given the name Draper. In time Draper II went into farming, married Mary Clifton, whose family roots were in the Nottinghamshire margins of the county, and set up home in Timberland. Halford's father, Draper III, was born in the village on May 4, 1818. For a time the farming activities of the family prospered. However, with the end of the Napoleonic Wars (1815), agricultural prices fell, and in the 1820s a run of bad harvests forced the family out of farming. Taking the remains of their capital, the Mackinders went to Derby, where they ran a public house without success. Then the Mackinders migrated to Liverpool to make a living in business, but this resulted in losses.[3]

Fortunately, arrangements had been made to apprentice Draper III to a physician in Bold Street, Liverpool, and he was able to start training, in his early teens, for a medical career. After Liverpool, Draper went to Glasgow, where he undertook training at Anderson's College and the Royal Infirmary. He witnessed some of the obstetrical work of Sir James Simpson, who introduced the use of chloroform into surgery. Draper later studied at St. Bartholomew's Hospital in London and became a member of the Royal College of Surgeons in 1847. In the following year he became a Licentiate of the Society of Apothecaries (LSA), which allowed him to practice medicine. He then

2. Halford J. Mackinder to C. A. Mackinder, c. July 1941, letter in possession of C. A. Mackinder, Edinburgh; Draper Mackinder farmer Rowston, Lease of Land, 1791, FANE 2/1/4/16, LAO.
3. Headstones in Rowston churchyard; Halford Mackinder to C. A. Mackinder; School of Geography, Oxford, Mackinder Papers, Autobiographical Fragments (M.P. Auto.).

took a post at Barton-under-Needwood in Staffordshire, working in association with William Birch, a former lecturer in midwifery at St. Bartholomew's.

It was probably during residence in Staffordshire that Draper Mackinder came into friendship with Dr. Halford Wotton Hewitt (1805–93), who ran a medical practice in Lichfield. Hewitt had trained in the old manner by apprenticeships and attendance at Stafford Hospital. It was not until 1845 that he had gained an M.D. at the University of Giessen in Germany. By the time Dr. Mackinder met Hewitt, the latter had a well-established family, with four sons and four daughters. Draper would marry one of the daughters, Fanny Anne, a decade later, but at this time he was still improving his medical skills and was not ready to settle down.

In 1851 Draper left Staffordshire to study in Paris. On his return he worked briefly at Oakham, in Rutland, but while in Paris he had met Dr. Septimus Lowe of Lincoln, who was working at the Hôpital la Piti. Septimus probably suggested that Draper go into partnership with a relative of his in Gainsborough. In 1852 Draper joined Dr. Francis Lowe at Elswitha Hall, a substantial Georgian house in the center of Gainsborough. Lowe's health was failing and, in spite of having a partner to share the work, he died within a few months of Draper's arrival.[4]

The practice, which Dr. Mackinder then ran singlehandedly, was not wealthy, but it did provide a living. However, rather than settling down to a quiet life in private practice, he continued his quest for qualifications. He prepared for the external M.D. examination at St. Andrew's University. In May 1853, along with thirty-five other candidates, Draper underwent several days of intensive examination work. He was one of the thirty-one candidates awarded the degree. More study followed, and in 1857 Draper became a Fellow of the Royal College of Surgeons, Edinburgh.[5] By then he was well qualified and ran a large practice in Gainsborough that served pri-

4. *London and Provincial Medical Directory* is the source of information relating to the medical appointments of Draper Mackinder, Halford Hewitt, and Septimus Lowe.
5. University Muniments, University of St. Andrews (I am grateful to

vate and pauper patients. He published in the leading journals and, in 1858, became one of the founding fellows of the
Obstetrical Society. In addition, Septimus Lowe and Draper
were instrumental in increasing awareness of public-health
problems in Lincolnshire. They contributed quarterly reports
to the *Journal of Public Health* on conditions in Lincoln and
Gainsborough. Draper became interested in medical geography and tried to relate the outbreak of diseases to environmental conditions. His contributions to the journal were accompanied by meteorological reports with an attempt to correlate
the incidence of disease to weather conditions.[6]

In 1860 Draper married Fanny Anne Hewitt. He was just
beyond forty years of age; his bride was twenty-eight. The following year a son, Halford John Mackinder, was born to the
couple. Although the Mackinders and the Hewitts lived on
opposite sides of England, they maintained close contact. There
were frequent Hewitt family gatherings to which Fanny Mackinder took her children. As a child Halford came to know
his uncle George Hewitt, who had a wide experience of southern Africa. On one memorable occasion, Halford met his Uncle
James, who lived in Columbia and traded on the Magdalena
River. Uncle Tom Hewitt also was able to convey strong impressions of other lands, for he had been educated at Darm-
stadt High School in Germany.[7]

Because the family of Draper Mackinder had been dispersed
as a result of the farming failure in Lincolnshire, it was the

---

the keeper of the Muniments, Robert N. Smart, for providing docu-
mentation); Records of the Royal College of Surgeons, Edinburgh.
6. *Sanitary Review and Journal of Public Health* 1 (1855). The book reviews in the first volume would bring Dr. Mackinder into contact
with the work of Dr. John Snow, *On the Mode of Communication of
Cholera* (London, 1855), and with Alfred Haviland's *Climate,
Weather, and Disease* (London, 1855). In volume 1 is an article by
A. K. Johnston, "Geographical Distribution of Health and Disease."
Johnston thought the "object of medical geography is to ascertain the
laws by which disease is distributed, and the manner in which conditions inimical to health . . . are found to prevail in certain regions
or localities." The laws, according to Johnston, depended on "the
facts of physical geography."
7. M.P. Auto.; *Who Was Who 1916–1928*, "Sir Thomas Hewitt" (London: Black, 1929).

Hewitts who provided Halford with grandparents, uncles, and aunts. Collectively they formed an important influence. Later in life, after Dr. Hewitt had served Lichfield as medical practitioner, sheriff, mayor, and justice of the peace, he moved to London and was able to subsidize his grandson, Halford, when he read for the bar.

Halford's home town of Gainsborough has been described as "one of the dreariest of the midland redbrick towns," but the first half of the nineteenth century was generally prosperous. The port of Gainsborough benefited from the increased river traffic generated by the growth of industry in the east Midlands. The shipbuilders of the town prospered. In 1815 a steam-driven vessel, the *John Bull*, was launched at Gainsborough, and the new vessel cut the journey time from Gainsborough to Hull from days to hours.[8]

Gainsborough was developing as an inland port serving the east Midlands and increasing its direct trade with the Baltic. By 1841 the foreign trade of the port warranted the opening of a customhouse.[9] All the town needed to link it fully into the economic life of the country was a railroad, and in 1849 the Manchester, Sheffield, and Lincolnshire Railway Company completed a line to Gainsborough. The line created new employment but turned out to be a mixed blessing. In general, railroads reduced the need for inland transshipment points such as Gainsborough, and when the line was extended to Grimsby, where the Royal Dock was being constructed (1849–52), the effect was to shift much of Gainsborough's trade to the Humber port. Between 1841 and 1861 the population of Grimsby rose from 3,700 to 15,000. At Gainsborough, after rising in the late 1840s, the population fell from 8,293 to 7,339 between 1851 and 1861. Eventually the railroad did facilitate the growth of engineering industries at Gainsborough, but the short-term impact was to cause a local depression. The position of the town in the world had been altered by the railroad. To what degree this local experience entered Halford's con-

8. Nikolaus Pevsner, *Lincolnshire;* Ian Beckwith, *History of Transport and Travel in Gainsborough.*
9. Reverend C. Moor, *History of Gainsborough,* p. 223.

sciousness is difficult to know, but an important segment of his Pivot paper was to be concerned with the manner in which railroads altered space relationships.

When the 1861 census was taken, the Mackinder household was modest, consisting of Draper and his wife, the new-born son, Halford, one live-in servant, and a medical assistant who did not have formal qualifications. However, the 1860s were prosperous for the family. Draper became local surgeon to the Manchester, Sheffield, and Lincolnshire Railway and to the Great Northern Railway. He began to work for the Marshalls, who were producing agricultural machinery and building up a workforce that exceeded two thousand by the end of the century. In 1869 Draper purchased Elswitha Hall from the heirs of Francis Lowe for one thousand pounds. By the census of 1871 other children (John, Fanny, Augustus, and Lionel) had arrived, and there were three servants living in, a governess of French nationality, Mme. M. H. Hosteller, who taught the children to converse easily in French, and Draper's medical assistant, a Dr. Henderson, who had recently taken an M.D. at Edinburgh.[10]

In the years that followed, Draper became the senior medical man in the town and the local medical officer of health. He continued to run his practice and found time to read widely and write short poems and devotional pieces.[11] After the travels of his younger days he was content to spend his time in Lincolnshire. He rarely left Gainsborough and did not take vacations.

Although Draper was a successful doctor, he was never affluent. The children were well educated, but expenditures had to be carefully watched. There were few extravagances, and the family never accumulated money to provide private in-

10. Census Enumerators Returns for Gainsborough, Enumeration District No. 4, pp. 34–35; Summary of deeds of Elswitha Hall, MCD 532, LAO; Census Enumerators Returns for Gainsborough, Enumeration District No. 4, p. 32.
11. Draper Mackinder lived to be ninety-seven. For his ninetieth birthday (May 4, 1908), his children had privately printed some selections from Draper's manuscript volume "My Recreation," which contained poems and other pieces. *The Gainsborough News*, May 22, 1908, reprinted several of the poems.

comes or buy partnerships in the professions or the business world. The ethos of the household was that you made your own way by hard work and intelligent application. Money was always tight, and Halford remembered the household being run in an economical manner.

Elswitha Hall was a solid three-storied Georgian house with bow windows that looked out onto a secluded garden at the rear. By contrast, the front of the house faced Clask Gate Street, which lay beside Gainsborough's River Trent waterfront. The narrow, cobbled street was a jumble of warehouses, wharf yards, workshops, and dwellings for laborers and the merchants who depended on the river for a living. Halford recalled thinking of the Hall as an island at the heart of a busy city. But from the nursery on the third floor, another view of the world opened out. To the west the floodplain of the Trent, stretching away into Nottinghamshire, displayed a land of villages and agricultural activity. For the son of a doctor the countryside was never far away. Halford often accompanied his father on the rounds that took them into the villages situated on the higher ground immediately to the east of Gainsborough or north onto the Trent floodplain and the carr lands around Haxey, where Draper was the district poor-law doctor. It is doubtful that they went farther north on the Isle of Axholme for the purpose of making a pilgrimage to Epworth, the birth place of John Wesley. Dr. Mackinder was Anglican, and nonconformists were not invited to dinner at Elswitha Hall.

Halford enjoyed outings to Lincoln, the cathedral town, where extensive Roman remains still formed a part of the fabric of the Victorian town. On occasion there were excursions beyond Lincoln, across the deeply cut valley of the River Witham, and south to Mere Hall.[12] The seventeenth-century hall stood on the Jurassic Ridge, or Lincoln Cliff as it is known locally. The Roman road, Ermine Street, followed the ridge from south to north through the county. In the grounds of Mere Hall were the relics of a hospital of the Order of St. John of Jerusalem, and close by were former properties of the Knights

12. M.P. Auto. Most of these early remembrances are taken from this source.

Templars. From the top of Lincoln Cliff there were fine views west toward the Trent and Newark. To the east were the fenlands, and it was a short trip to see the village of Rowston, where the ancestral Mackinders had farmed, and Timberland, where Draper had been born into rural England a half-century before.

When Halford went to school at Epsom College in 1874, he met Thomas Walker (1861–1945), who lived at Hundleby House, Spilsby, at the southern end of the Lincolnshire Wolds.[13] The boys were the same age, became firm friends, and eventually went up to Oxford together. They probably visited each other in the vacations: exchanges that would have allowed Halford to see the chalk Wolds with their thin soils, large farms, and modest villages. We do not know whether he visited the tiny village of Somersby, just north of Spilsby, where Tennyson (1809–92) had been brought up in the rectory. But the works of the poet laureate probably were read at Elswitha. Draper Mackinder liked to compose poems and attempted to reflect Tennyson when he wrote such lines as:

> All onward moving, slow or fast,
> should eye the future through the past.[14]

Halford was fascinated by the countryside and at the age of ten was given a pony on which he rode into the surrounding area. The years between nine and sixteen were a rich time of imagining, speculating, and being influenced first by the landscapes of Lincolnshire and then by the Downs around his boarding school at Epsom. In later life Mackinder could recall vividly the scenery of his part of Lincolnshire: floods on the Trent, the windmills on the higher ground to the east of Gainsborough, and the aegre, or tidal bore, on the river. He remembered:

> We children were allowed to sit up when the moon was at full and a more than ordinary aegre was expected. We used to be taken down to the steam packet wharf and after a hushed wait,

13. *Who Was Who 1941–1950*, "Sir Thomas Walker" (London: Black, 1952).
14. "My Recreation." See n. 9.

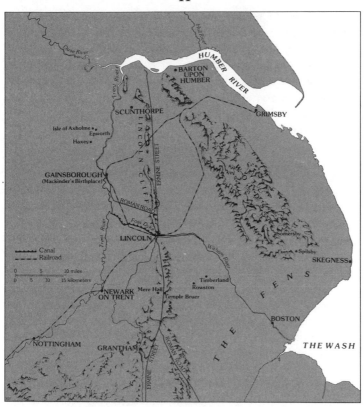

Lincolnshire

broken by the clanking of chains as the barges at anchor prepared for the rush of waters, the distant roar would be heard.
The bargees sang out "war Aegre" and the cry was repeated up
the river. Presently . . . a front of waves several feet high . . .
came up swiftly.[15]

Much of Mackinder's writing expressed broad views of the
world scene, but his geographical knowledge was firmly rooted
in the landscapes of the English countryside. Throughout life
he was a keen walker. When there was a large job of writing

15. M.P. Auto.

to be completed, he usually got away to some quiet farmhouse where he could walk, think, and speculate alone. It would be a mistake, however, to believe that he had little firsthand knowledge of the urban-industrial scene.[16] Lincolnshire does not possess much industry, but it did go through a phase of industrialization in the second half of the nineteenth century, when Mackinder was growing up. Gainsborough, an outlier of industrial Nottinghamshire in rural Lincolnshire, displayed many of the characteristics of a northern manufacturing town. At both Gainsborough and Lincoln there were large engineering works, and at Scunthorpe iron and steel works were developed during his boyhood.

The central part of Gainsborough, where the Mackinders lived, was almost entirely working class. Living conditions were poor, and Draper Mackinder and his family had daily contact with the health problems of the area. Many large Georgian houses had their back gardens "redeveloped" in the form of courts containing rows of cheap cottages. In one of his reports to the *Journal of Public Health*, Dr. Mackinder described these courts as narrow, ill-ventilated, and offensive. The health problems were partly the result of bad building practice, but Dr. Mackinder thought the residents made matters worse with a form of "atmospherophobia—a horror of fresh air, a dread of draughts."[17] Dr. Mackinder had a wide experience of medical problems in industrial areas and told arresting stories of his work with cholera epidemics in Liverpool (1832), London (1847), and Gainsborough (1866).

Although Gainsborough was an industrial town, it served as a market for the surrounding country and had many rural attributes. Not far from the working-class area where the Mackinders lived was another social system headed by the principal landowner, Hickman Bacon. Here was a society composed

16. In *The Mill on the Floss* (1860), George Eliot used Gainsborough as a model for her town St. Ogg's. Eliot portrays well the urban and rural character of Gainsborough. The life of Draper Mackinder contains many facets that fit the training of Dr. Lydgate, a major character in Eliot's *Middlemarch* (1871–72).
17. *Sanitary Review and Journal of Public Health* 3, (1857). Report for September, October, November, 1857.

of landowners, substantial farmers, and professional families. As a doctor, Draper Mackinder had access to this world, but leading businessmen, such as the Marshalls, who manufactured agricultural equipment, were far from being accepted. Eventually Henry Marshall, a founder of the works, became a justice of the peace and played a part in national politics as a member of the tariff-reform movement.[18] The social division of Gainsborough was not without influence upon Halford. In later life he recalled his ambivalent attitude towards the Marshalls, whom he admired yet treated with reserve. Had he been able to ally himself with the Marshalls, with their commitment to tariff reform, his political career might have been more effective.

At Elswitha Hall Halford was given instruction by the governess, Mme. Hosteller, and his command of French grew to the point where he could think, visualize, and dream in a second language. At the age of nine he started to attend the Gainsborough Queen Elizabeth Grammar School (founded in 1589), which then occupied a large Georgian house at Hickman Hill. He did exceptionally well at scripture, history, and geography, and he enjoyed drawing maps. He took an early interest in world affairs and was fascinated by the Franco-Prussian War (1870–71). The war must have been of great interest to a family in which the father had experience of France, the children had a French governess, and the mother had a father and at least one brother who had been educated in Germany. Mackinder's suspicion of German expansion may well date from this period of his life.

Of Halford's boyhood reading we know something. In the notes he made in later life he did not mention the nature books and historical romances of the Gainsborough writer Thomas Miller (1807–74), but he did remember his excitement in the library of the Gainsborough Mechanics Institute as he read about Captain Cook's voyages. He recalled his mother reading the adventures of Peter Parley to the children. Peter Parley is a forgotten character, but in the Victorian nursery he was popu-

---

18. *Who Was Who, 1897–1915*, "Henry D. Marshall" (London: Black, 1920).

lar, and the books went into edition after edition. The volumes, written by an American publisher, Samuel Griswold Goodrich (1793–1860), covered a great range of topics. There were books on the *Geography of the Bible, Tales about Universal History on the Basis of Geography, Peter Parley's Wonders of the Earth, Sea and Sky,* and *Tales about the Sea, and the Islands in the Pacific Ocean.* In all there were more than one hundred titles. The books imparted factual knowledge within the context of adventure stories. The technique was successful, and the titles sold millions of copies in the mid- and late nineteenth century. Probably the books formed an important element in the early education of young men and women who subsequently helped run the British Empire or open the American West. Not all the books were written by Goodrich. For example, Nathaniel Hawthorne played a large part in the compilation of Peter Parley's *Universal History on the Basis of Geography,* and many volumes were substantially edited for the English market, particularly by clergymen.

We shall never know how many Peter Parley volumes Halford heard in the nursery, but he remembered the adventure stories clearly. They must have helped create a vision of the wider world that his Hewitt grandfather and uncles reinforced. In addition to the adventure stories and potted descriptions of history and geography, there were Parley atlases. It may have been from these that Halford began to draw maps, an activity he greatly enjoyed in boyhood. Any reasonable acquaintance with the Parley material would have given Mackinder a grasp of what has been called "capes and bays geography." Thus, when in 1887, speaking on the scope and methods of geography, he suggested that a traditional treatment of southeast England would start with a listing of the capes and inlets of the coast and proceed with an enumeration of the major valleys, he was speaking from a firsthand knowledge of the geography that had been put in front of young people in his childhood.

If we look into the Mackinder household for intellectual influences that may have shaped Halford's geography it is easy to recognize the international elements. The Hewitts were travelers who enjoyed living overseas, and Halford Hewitt had a particular interest in Germany. Draper Mackinder, impressed by his visit to France in the 1850s, had engaged a French governess to instruct his children. Of course there was nothing special in Halford's having well-traveled family connections. Many middle- and professional-class households in Victorian Britain had relatives serving in regiments, consular offices, and professional posts around the globe.

Draper Mackinder was an important intellectual influence. He was broadly read and well trained as a scientist. Perhaps most important of all, he had a strong interest in medical geography. From this parental influence Halford was taught to look for interrelationships between factors in the environment and was encouraged to think of distribution as a key to possible causes.

The landscapes of Lincolnshire contributed to Halford's intellectual growth. The county has easily recognizable regions — the Wolds, the Fens, the Jurassic Ridge, the Isle of Axholme — in which farming activities reflect changing physical conditions. The historical record of the county sits plainly in the landscape. The Roman roads and towns stand out, and the work of later drainers and enclosers is broadly evident. Some of the more incisive parts of Mackinder's geographical analysis concern his understanding of the differing value of locations and the way location can be altered by technological means. Just before and during his boyhood the railroads were altering the value of locations in Lincolnshire as the county experienced late industrialization.

By the age of thirteen Mackinder had acquired many interests that he would explore more fully later. In November 1873, at the Old Hall, he made presentations on Australia and on ivory and in so doing discovered his enjoyment of lecturing to an audience. At this time his parents made the decision to send him away to school because the small town of Gainsborough lacked a full range of educational opportunities.

2
Epsom and Oxford

Draper Mackinder wanted his son Halford to be a doctor. However, Gainsborough, like most small towns in England, did not have a proper system of secondary education. If Halford were to prepare for entry to medical school, he would have to go away to school. Because a number of professions, including medicine, were beginning to require a degree, in the second half of the nineteenth century there was an increasing demand for schools to prepare students for university. There being no national system of secondary education, a large number of new, fee-paying schools were built to complement the few existing institutions such as Eton, Harrow, Rugby, Winchester, and Westminster.

Epsom College, where Halford received his secondary education, was a new school, opened in 1855 under the auspices of the Royal Medical Benevolent College as a result of initiatives by John Propert (1793–1867), a successful London surgeon. At first the school aspired to educate two hundred sons of medical men at relatively low fees, even offering forty free places. The early history of the school, however, was unhappy. The staff-student ratio was poor, and academic standards were low. There were complaints about the care of the boys, and the curriculum lacked purpose.[1] The school day was filled with religious instruction, classics, and rote learning exercises in history and geography. Curricula of this type were common,

1. Michael A. Salmon, *Epsom College 1855–1980*. In 1857 and 1858 a number of letters in *The Lancet* questioned the governance and management of the Royal Medical Benevolent College, as the school was then called. "Paterfamilias" closed his letter, in the issue of January 16, 1858, with the slogan "less work and more play; a better diet, and fewer floggings."

but it was surprising at Epsom, given the medical background of founders and pupils.[2]

Epsom College established a sound reputation, under the headmastership of the Reverend William de Lancy West, between 1870 and 1885. West was a classicist, but he wasted no time in bringing science into the curriculum. He made a number of part-time science appointments, one going to William Thiselton Dyer (1843–1928), an innovative botany teacher. Dyer had a natural-science degree from Oxford and knew leading biologists such as Henry Moseley, Ray Lankester, and Thomas Huxley. He would become director of Kew Gardens. By the mid-1870s, when Halford arrived at Epsom, the curriculum included physics, chemistry, botany, and zoology. Much of the work was practical, instruction was given in a laboratory, and students were made familiar with microscopes. In the establishment of the sciences Epsom was ahead of many schools.[3]

Acting as Gainsborough's local secretary and treasurer to the Royal Medical Benevolent College, Draper had little difficulty choosing Epsom for his son. In September 1874 Halford went south to enter Epsom College as a pupil. The train journey through south London must have been fascinating for the young schoolboy from Lincolnshire, but soon the line cleared the built-up area, running among farms and fields (since covered by the expansion of the city). The railway cut deeply into the North Downs, exposing the chalk. The line terminated on the dip slope of the Downs, far from any built-up area. Halford would have walked out of the Epsom Downs station and looked at the grassy slopes of the unenclosed downs. To the left, up and over the hill, was the Epsom race course; to the right was the red brick mass of Epsom College, which even today looks "gothic and grim."[4] Straight ahead, a mile to the west, was the small market town of Epsom. The college, new, raw, and red, stood out in formidable isolation.

The school and the surrounding countryside made a strong

2. Report of the Education Committee, December 15, 1865, Records of Epsom College (REC).
3. Minutes of Education Committee, February 8, 1872, February 21, 1872, April 1, 1873, June 5, 1874, July 7, 1874, REC.
4. Ian Nairn and Nikolaus Pevsner, *Surrey*.

impression upon Halford. The small town of Epsom, with its weatherboard houses and superb Queen Anne and Georgian buildings, had a style and graciousness that must have given it the air of a foreign land, compared with Gainsborough. If there were thoughts that similarities existed between the North Downs and the Lincolnshire Wolds, they would be quickly dispelled on contact with the surprisingly steep slopes and incised valleys of the Downs, which hosted extensive beech woods and thickets of hawthorn. The soils were thin, as on the Wolds, but there was little arable farming on the high Downs, which were bleak in cold or wet weather.

During his years at the school (1874–80) Halford enlarged his knowledge of Surrey, until he felt he had tramped all over the county. He knew the Downs and the Weald well and wrote an article on the geology of the area around Epsom College for the school magazine. He read William Topley's memoir, *The Geology of the Weald,* a work that made important contributions to an understanding both of the Weald river system and of the relationship between settlement patterns and geological features. Mackinder's interest in geomorphology, well displayed in *Britain and the British Seas* (1902), probably stems from his reading of Topley's memoir. Mackinder built on Topley's ideas concerning the evolution of settlement when he came into contact with the work of the historian J. R. Green at Oxford in the early 1880s. This interest in the history of settlement is evident in the "Scope and Methods" paper and in the lectures on historical geography that Mackinder gave in Oxford from 1887 to 1905.[5]

Halford was an active schoolboy. He took part in school plays and the debating society, and he won a place on the rugby team. At Founder's Day each year he took more than his fair share of prizes for English, French, and drawing and for ranking at

5. H. J. Mackinder, "Geological Epsom," *The Epsomian* 10 (1880); W. Topley, *The Geology of the Weald;* S. W. Wooldridge and Frederick Goldring, *The Weald,* p. 202; W. Topley, "On the relation of the parish boundaries of South East England to Great Physical Features," *Journal of the Anthropological Institute* 3 (1872). R. V. Lennard's notes taken at Mackinder's lectures on historical geography at Oxford are in the Bodleian Library, Ms. Top Oxon e. 419.

the top of his class. He made contributions to the school magazine on geology and zoology and at age sixteen wrote a fascinating piece, "A Glimpse of A.D. 1950," in which high-speed electric trains, using a channel tunnel and a suspension bridge over the Bosphorus, linked Britain and India in a journey time of sixty hours. The prognosis was not accurate, but the article does indicate an early interest in the long-term pattern of events.[6]

Mackinder was only one of a bright batch of students that West's Epsom was nurturing. The two Lincolnshire boys, Walker and Mackinder, were close friends and keen rivals. Thomas Walker was better at rugby, mathematics, the hard sciences, and debating. Mackinder's strengths were essay writing, languages, environmental sciences, and public speaking. At debates the two boys were usually found on opposite sides, and Walker seemed to come out the winner more frequently, perhaps because he chose his debating ground more skillfully. When the boys sat for the London University matriculation examination in June 1878, both did well, but Mackinder was higher on the list and won Epsom's Gilchrist scholarship, which would have given him £50 annually toward the cost of a medical education at St. Bartholomew's medical school. The next year they took the preliminary science examination at London University. Walker passed, but Mackinder failed, as he did in the summer of 1880. Halford passed the examination the following summer, when he was an undergraduate at Oxford. In their last year at school, 1879–80, Halford and Thomas tried for awards at Oxford. In November Mackinder won an exhibition in physical science at Merton College, worth £40 a year, but bigger prizes were to come. In February each man was offered a Junior Studentship at Christ Church, valued at £85 a year. The majority of Junior Studentships was reserved for pupils from Westminster School; only a handful were open to others. For Epsom to win two awards indicated that science teaching at the school was first rate.[7]

6. H. J. Mackinder, "A Glimpse of A.D. 1950," *The Epsomian* 7 (1877).
7. Founder's Day Program: Prizes 1880, REC; M. P. Auto.; *The Epsomian* 9 (1879); Junior Studentship, 1850–1880, 39, B.b, Christ Church, Oxford.

Among the prizes each year at Epsom was a Propert Medal, awarded to the outstanding boy. In the summer of 1880 two Propert Medals had to be produced—one for Mackinder and another for Walker. The boys collected many other prizes at Founder's Day, but Halford's mother had difficulty enjoying the occasion. Draper Mackinder, on one of his rare trips away from Gainsborough, was overcome with shyness in Epsom and was unable to enter a restaurant to order lunch. His wife suffered acute pangs of hunger. What Draper thought of Epsom College is not recorded, but he sent three sons—Halford, Augustus, and Lionel—to the school expecting that at least one would become a doctor. The college produced an academic, an insurance assessor, and a cleric. It would be unfair to blame the school. Draper Mackinder had qualified himself massively as a doctor prior to devoting his life to the service of an obscure town in the east Midlands. The sons must have sensed that it would be difficult to match the professional attainments of their father in medicine.[8]

In October 1880 Mackinder and Thomas Walker went up to Oxford and entered Christ Church. They were housed in the Meadow Building overlooking Christ Church meadow and the River Isis. Once more Mackinder was beside navigable water and a broad river valley, although there were differences between the working vessels on the Trent and the pleasure boats on the upper reaches of the Thames.

Mackinder's tutor was the Reverend Francis Paget (1851–1911), a future dean of Christ Church and bishop of Oxford. Paget, who was High Church, felt the need to instruct his young, provincial charge in his duty as a churchgoer. This Halford resented. He came from a family that knew its religious duties, and he moved in groups that contained young men who were to take Holy Orders. Later, Dean Paget and Mackinder would become firm allies in the foundation of Reading University.[9]

Both Walker and Mackinder took science degrees. Thomas

8. Founder's Day Program: Prizes 1880, REC; *Epsom College Register October 1855 to July 1905*, REC.
9. Tutors Book, M.T. 1880, Christ Church, Oxford; M.P. Auto.

specialized in physics and mathematics. Halford specialized in animal morphology, but he also took examinations in physics, chemistry, physiology, and botany. Mackinder immersed himself in a wide and demanding range of university activities. Most of his afternoons were spent in the laboratories, at the University Museum, which meant that much reading had to be done in the evenings. Nevertheless, he joined the Union, helped found the Junior Scientific Club in 1882,[10] and enlisted in the Oxford University Rifle Volunteers, being promoted to lieutenant in 1883. Subsequently he took courses at Wellington Barracks and Aldershot and passed through the field-officer school. For a time he intended to join the Royal Engineers, but the idea faded as his academic interests grew.

It was not easy to offer the army a new idea, but Mackinder tried. Having conceived a scheme to establish university officer-training corps, he wrote a proposal, which his commanding officer rejected. (The idea reemerged, in the early years of the twentieth century, when university corps were set up.) The connection with the army was not to be wasted time in the training of a geographer. During the summer vacations, Mackinder took part in exercises that involved long marches across the countryside. He got to know the Hampshire Basin and the south coast well, just as he already knew the London Basin and the Weald.[11]

Another military interest was the war-game club, founded some years earlier, when Spenser Wilkinson, the military historian, was an undergraduate. In Mackinder's time the club was led by two dons in the Rifle Volunteers: Harry Reichel of All Souls and H. B. George of New College. Reichel, Mackinder's sergeant in the Volunteers, later became Sir Harry, vice-chancellor of the University of Wales. Herford George had an influence upon Mackinder. He pioneered the teaching of mili-

10. Oxford University Junior Scientific Club, Minute Book 1, Radcliffe Science Library. The club was established at a preliminary meeting in Trinity College on November 18, 1882. Mackinder was elected to the committee. The first regular meeting was held on December 1, 1882. I am indebted to Paul Rowlinson and Professor Margaret Gowing for help with these records.
11. M.P. Auto.

tary history at Oxford and wrote on the relationships between history and geography and on the historical geography of the British Empire. George proposed Mackinder for membership of the Royal Geographical Society in March 1886.[12]

At the Oxford Union Society, Mackinder spoke in a high percentage of the debates while he was an undergraduate, and he was elected secretary, treasurer, and president. During his term as treasurer, electric lighting was installed in the Union, which gave rise to debating barbs about the treasurer and his "electric friends."[13]

At the Union Mackinder became part of a group of undergraduates who were to have a powerful influence on the growth of Oxford University Extension teaching and, eventually, on educational reform. The group revolved around Michael Sadler (Trinity College), and included Hudson Shaw and Gordon Cosmo Lang, both of Balliol. The young men had a set of definable characteristics. They came from middle- and professional-class backgrounds in northern provincial areas, they made their way by scholarships, and their ambitions for reform carried with them a strong sense of moral purpose. They came from families with traditions of religious commitment, community service, and reform activity. Several, including Shaw and Lang, went into the Church; Lang became successively archbishop of York and archbishop of Canterbury.

If Mackinder's ideas about modernizing English education were taking shape at the Union, it was at the Science Museum that his interests in biology, geology, and world affairs began to grow, fuse, and create the new perspectives out of which geography would emerge as an academic subject in England. In his second year Mackinder was tutored by the Lee's Reader in Anatomy, John Barclay Thompson. Thompson's research in-

12. Sir Charles Oman, *Memories of Victorian Oxford*, pp. 108–109, describes the University Kriegspiel Club around the time Mackinder was a member. Henry Spenser Wilkinson, *Thirty-Five Years*, p. 7, contains information on the origins of the club. Mackinder was elected to the Royal Geographical Society on March 22, 1886. H. B. George, Douglas Freshfield, and Reginald Cocks recommended admission.

13. Oxford Union Society, Rough Minute Book, December 1876–January 1884.

terest was in the distribution of present and extinct species of fish. The major academic influence, however, was Henry Nottidge Moseley (1844–91), Linacre Professor of Human and Comparative Anatomy. Moseley had a wide training in science, having studied at Oxford, London, Vienna, and Leipzig. He had been on a scientific expedition to Ceylon and was then attached to the staff of the *Challenger* expedition between 1873 and 1876. Moseley was appointed to *Challenger* as the zoologist, but he also served as botanist and ethnographer. Before joining *Challenger*, Moseley went to see Charles Darwin (1809–82) to seek advice. Darwin suggested that Moseley collect materials relating to cultures in remote areas, because missionaries and commerce were rapidly destroying traditional life styles. Moseley returned from the *Challenger* voyage with a large collection of zoological and botanical specimens, together with a mass of ethnographic material. He had also acquired a share in a guano deposit that yielded a substantial income.[14]

The year after returning from *Challenger*, Moseley visited California, Oregon, and Washington and wrote a book assessing the natural resources and indigenous cultures of the Far West.[15] After work at London University he returned to Oxford in 1881 to reorganize the teaching program in animal morphology. He gave long, detailed lectures to his students and spent many hours organizing dissecting work in the laboratory.

The group of students around Mackinder included: Henry Balfour (1863–1939), who went on to run the Pitt Rivers Museum; Baldwin Spencer (1860–1929), who became professor of zoology at Melbourne and undertook definitive studies of Australian aboriginal cultures; and G. C. Bourne, who succeeded to the Linacre Chair in 1906. The students lived in an

14. Tutors Book, M.T. 1881, Christ Church, Oxford; Peter Chalmers Mitchell, *My Fill of Days*, pp. 77–78; Henry Nottidge Moseley, *Notes by a Naturalist; An account of observations made during the voyage of H.M.S. "Challenger" round the world in the years 1872–1876*. The 1892 edition of Moseley's book carried a biographical account of him written by G. C. Bourne. Originally the book was published in London, by Macmillan, in 1879.
15. H. N. Moseley, *Oregon: Its Resources, Climate, People, and Productions*.

atmosphere of constant discovery as studies in comparative morphology provided the detailed work supporting the Darwinian view. Mackinder performed well within the group and was awarded a first-class honors degree in 1883. Twenty-one students took the animal morphology examinations, and six were awarded "firsts."

It is not surprising that Moseley produced some of the leading zoologists of the day, but for important figures in geography (Halford Mackinder) and anthropology (Baldwin Spencer) to emerge from the same body of students tells us much about the breadth of his interests and influence.

Moseley was part of a British scientific and exploratory tradition that stretched back at least into the eighteenth century when Captain James Cook sailed the *Endeavour* (1768) into the Pacific to observe the transit of Venus, and the scientists on board, including Sir Joseph Banks, collected data and specimens that enlarged English knowledge of the world and its resources. The tradition strengthened in the nineteenth century with the admiralty hydrographic surveys and associated work. The voyage of *The Beagle* (1831–36) and Charles Darwin's studies, leading to a greater understanding of evolutionary processes, were a part of the tradition, as was Thomas Huxley's work on HMS *Rattlesnake* (1846–50). The *Challenger* expedition on which Moseley served, the most elaborate of the surveys, collected a mass of biological, climatic, and oceanographic data that facilitated the development of a more complete global view of the physical environment.

Moseley became a key figure in the effort to establish the teaching of geography at Oxford. In 1881 he was elected to the Royal Geographical Society and took on the task of assessing the papers in physical geography in connection with the RGS school essay prize that existed from 1869 to 1884. Moseley began to press the view upon the RGS that more could be done to advance the teaching of geography. In 1884 his examiner's report suggested:

> The more I gain experience as an examiner in physical geography the more I am convinced of its value as a subject of general education and the more I regret that it is not represented

in Britain as it is in continental universities. It is almost hopeless to expect that schools will do it justice until it is recognized at the universities and scholarships and other inducements are offered to those more proficient in it.

This view was already established in some quarters of the RGS, and early in 1884 the scientific-purposes committee began to urge that more be done to promote the place of geography in education. One member of the committee, Douglas Freshfield, proposed that Oxford be offered £300 annually for five years in order to appoint a professor of geography. It was even suggested that Moseley be invited to take up the new chair.[16]

When Mackinder took his natural-science degree in 1883, he and his friends were not keen to leave Oxford and take up traditional careers in school teaching, law, the army, or the civil service. Thomas Walker did go back to Epsom College to teach for a while, but most of Halford's friends were interested in doing more academic work and developing ideas that would broaden English education. English universities were in an expanding phase, and new subject areas were developing. Provincial university colleges had been founded at Newcastle (1871), Bristol (1874), Sheffield (1879), Birmingham (1880), and Nottingham (1881). The Victoria University, serving Manchester, Leeds, and Liverpool, had been chartered in 1880.

In Oxford numerous changes were taking place. An act of Parliament (1877), which had provided for the reform of the University, resulted in the appointment in 1882 of commissioners who oversaw the establishment of a common university fund, the creation of several new faculties, and the reorganization of teaching with the readers and professors taking a more active part in lecturing. These changes opened up possibilities for establishing new posts and teaching new subjects. The intellectual development of the university was reflected in the creation of new buildings and institutions: Examination Schools (1882), the Indian Institute (1884), Pusey House

16. Examiners Report in Physical Geography for the RGS Prize 1884, Moseley file, Royal Geographical Society (RGS); Committee Minute Book, March 1883–December 1890, Scientific Purposes Committee, February 8, 1884, February 15, 1884, RGS.

(1884), the Pitt Rivers Museum (1887), and the foundation of women's colleges. Lady Margaret Hall opened in 1878, Somerville in 1879, and St. Hugh's in 1886.

The men in Mackinder's circle were attracted by these developments and must have sensed that they would be given opportunities to try new ideas if they could stay in Oxford and broaden their training. Money was not a problem; they either had scholarships or won new awards. Mackinder's junior studentship, at Christ Church, ran for five years so long as he read for a degree. But which degree? At this time there were no higher degrees at Oxford of the type offered at German universities and which American universities were introducing.

In the academic year 1883–84 Mackinder read for a history degree in order to see "how the theory of evolution would appear in human development."[17] He heard A. L. Smith (1850–1924) lecture on constitutional law, and he read J. R. Green's *The Making of England*, which influenced his thinking about historical geography. He attended the lectures of Bonamy Price (1807–88) on political economy and learned about currency and banking. Professor Price advocated the teaching of social sciences at universities, and his views impressed themselves on Mackinder and his circle.

Halford read heavily in geology. What he read reinforced an existing interest and developed the knowledge of paleontology that he had acquired in Moseley's classes. Early in 1884 Mackinder competed, by examination, for the Burdett Coutts Prize in geology. The prize was a postgraduate fellowship, renewable for two years, established to promote research in geology. The examiners usually included questions on the geology of the Oxford region and Halford added further to his detailed knowledge of southern England. Mackinder won the Burdett Coutts Prize, but his study of history slackened and he achieved a second-class degree in that subject in the summer of 1884.

With the Burdett Coutts Prize money added to his studentship, Mackinder was relatively well off, and in July 1884 he and Baldwin Spencer went for a long walking holiday in Europe. Traveling first to Strasbourg, they went on foot through

17. M.P. Auto.

the Black Forest and then into Switzerland. When their money ran out, they returned to London. The disciplines of anthropology and geography probably owe much to the young men's observations and conversations during this long walk. The experience also gave them confidence for the expeditions they were to launch in the future. Mackinder was to climb Mount Kenya, and Baldwin Spencer was to take lengthy walkabouts in the interior of Australia to study aboriginal culture.

When the 1884–85 academic year started in Oxford, Mackinder began to read for a law degree, but he did not neglect his other interests. He was elected president of the Junior Scientific Club,[18] while Sadler was working to establish the Oxford Economic Society. Whether or not Mackinder ever took the research in geology seriously, which the Burdett Coutts scholarship required, cannot be documented. However, around this time he made a careful study of geomorphology, and the results are apparent in the 1887 "Scope and Methods" paper and later in *Britain and the British Seas* (1902).

In the spring of 1885, Michael Sadler and his friends had a stroke of luck. Arthur Acland, secretary to the committee for Oxford University Extension, left to pursue a political career in London, and Sadler was invited to take over Acland's position. Here was an opportunity to try out some of the educational ideas that Sadler had developed. Sadler quickly expanded the extension program and brought in many new lecturers, including Mackinder, Lang, Hudson Shaw, and J. A. R. Marriott (1859–1945).[19] Remuneration to the lecturers was not great, but a single man could manage on it for a time. The major problem was that there was no security and little opportunity to develop a career as an extension lecturer.

Mackinder hedged his bets. He agreed to lecture for Sadler

18. Oxford University Junior Scientific Club, Minute Book, June 6, 1886–December 5, 1889, Radcliffe Science Library. The first meeting with Mackinder as president took place on October 24, 1884. Baldwin Spencer read a paper on the Rhine Valley.
19. Committee for University Extension, Minute Book, June 6, 1885–December 5, 1889, Rewley House, Oxford (RHO). Mackinder, Oliver Elton, and C. W. C. Oman were approved as extension lecturers on June 6, 1885. I am indebted to Mr. A. J. Trump, librarian, for help with the Oxford extension records held at Rewley House.

but left his law studies in Oxford and went with Thomas Walker to London, to read for the bar and qualify as a barrister. Before moving to London, Halford talked the situation over with his grandfather, Halford Hewitt, then resident in St. John's Wood, and with an uncle, Thomas Hewitt, a barrister specializing in tax law. Halford Hewitt lent his grandson a hundred pounds to enter the chambers of Gorrell Barnes, a leading shipping lawyer. Mackinder got on well with Barnes, dined at his home, and even persuaded the lawyer to become a member of the Royal Geographical Society.[20]

During 1885–86 Michael Sadler started to transform Oxford University Extension work from a polite expression of interest in improving the education of those not at university to a crusade to bring knowledge to the working man. Sadler took on new lecturers, new courses, and new centers. He was indefatigable in the production of circulars and in contacting institutions in towns where he thought there might be a demand for extension work. He produced a brochure, *Co-operative Societies and Oxford University Extension Lectures*, and looked north, to the provincial towns from whence he and his associates had come, for potential locations for lecture series. In towns such as Ilkley, Barnsley (his birthplace), Doncaster, Rotherham, Gainsborough, and Lincoln, Sadler was successful in establishing new lecture programs. On an early venture, late in 1885, Mackinder and Sadler went to the Rotherham Mechanics Institute to give a short series of special lectures under the auspices of the Masboro Co-operative Society. Mackinder talked to an audience of four hundred about physical geography and the wonders of the scientific method, and Sadler spoke on economics.[21] Frequently the early lectures were on economic theory and touched upon the history of the cooperative and trade-union movements.

Early in 1886 Sadler issued his Circular 16, which listed

20. Certificate of Candidate for Election, Frederic Gorell Barnes, elected January 31, 1887, RGS. The certificate was signed by Mackinder and Keltie.
21. *Rotherham Advertiser*, December 5, 1885, February 6, 1886, February 13, 1886. Quoted by Alice Garnett: "Climatology and Urban Geography," *Institute of British Geographers, Transactions* 42 (1967): 21–22.

the courses available for the 1886–87 academic year. For the
first time geography was listed as a separate subject area, with
courses offered in physical and political (human) geography.
By this time Canon A. G. Butler, fellow of Oriel and relative
of Francis Galton, was on the extension committee, and he
contacted the Royal Geographical Society to suggest that the
society encourage the geography lecture program by the pay-
ment of a small subsidy to Sadler's office.[22]

As far as Sadler was concerned, the academic year 1885–86
was largely spent laying plans and setting up a sound administrative structure. The number of lectures delivered was still
modest. From Mackinder's viewpoint this had advantages, for
it allowed him to concentrate on studying for the bar. In June
1886 Walker and he sat their bar finals and were called at the
Inner Temple.

In the fall of 1886 Mackinder pursued two careers: barrister and university extension lecturer. Walker and Mackinder
set up at 6 King's Bench Walk and began to receive briefs on
the Midland Circuit. In one case Mackinder created a stir. A
Gainsborough solicitor retained him to defend a client against
whom the police had assembled evidence.[23] The solicitor felt
that the evidence was spurious, and Mackinder destroyed the
police case in court. It must have been a formidable experience to be cross-examined by the former president of the Oxford Union, now a newly minted lawyer, whom Henry Moseley
had trained at the Science Museum to believe nothing that
could not be fully verified. No doubt the police witnesses were
glad to get out of the court.

Mackinder could have made a successful career at the bar.
He had an unusual capacity for mastering new materials, fine
expository powers, and a presence that would have exposed
uncertain witnesses. For Mackinder, however, the development
of new subject areas and new forms of higher education was
a more exciting mission—and to Mackinder it was a mission.
In 1886–87 he was extension lecturing for Sadler and working the Midland Circuit as a barrister. Having spoken both in

22. Committee for University Extension, March 3, 1886, March 11,
    1887, RHO.
23. M.P. Auto.

courts and to audiences of earnest people keen to improve their education, he had little difficulty in choosing the educational career. Mackinder was a man of conviction. He would not have been easy arguing cases when he was uncertain his side was in the right.

Events were beginning to gain a momentum of their own. In 1886–87, after a year of planning, Sadler had a large, far-flung lecture program. Between November 1886 and March 1887 Mackinder gave courses of lectures, usually six presentations in the series, at Bath (2), Taunton, Bridgewater, Manchester (2), and Salisbury (2).[24] Not all the courses ran simultaneously, but those that did required long hours of travel. Sadler had the extension lecture program tightly organized. The speaker delivered a prepared lecture to the class and, after a short break, gave a tutorial. The students were provided with a printed syllabus that outlined the lectures and contained a reading list. To make books on the reading list available locally, Sadler started small, traveling libraries.

At the end of the course registered students were examined and the papers assessed by an outside examiner. In 1886, for example, Sadler engaged Henry Bates, assistant secretary of the Royal Geographical Society, to examine work in connection with the physical-geography courses Mackinder was offering. "You will, of course, understand that the examination is a check upon the lecturer as well as the student," Sadler told Bates and asked that he make suggestions for improvements.[25] Bates must have wondered at his wisdom in accepting the job, for he was paid a standard fee regardless of the number of scripts. The following April 27, Sadler wrote to Bates apologizing for the large number of examination papers. The fact that there were "exceptional numbers wishing to be examined, shows the wonderful interest taken in geography, in consequence of Mackinder's treatment of the subject," and would he get the papers back as soon as possible.

Bates's reports on the examinations agreed that the students

24. Oxford University Extension Lecturers and Examiners Reports, 1886–87, RHO.
25. Committee for University Extension, December 2, 1886, RHO; Sadler to Bates, April 27, 1887, RGS.

had profited from the lectures of Mackinder. Those students who passed the examination were given a certificate. However, it was of little value, serving neither as a substitute for matriculation (university entrance) nor for credit toward a degree at an institution such as Mason College, Birmingham. This was the great flaw in the extension programs offered by Oxford, Cambridge, and London universities, and nothing was ever done to rectify the problem. Consequently, after a burst of interest in the late 1880s and 1890s, the extension movement declined to become what it is today: an authoritative way in which relatively well educated people can develop their interests and hobbies. When the lecturers realized this, they moved on to other things. Some, such as R. G. Moulton of Cambridge, went to the University of Chicago and other colleges in the United States, where extension work could count toward degrees. Michael Sadler left extension work at Oxford in 1895. He became director of the newly created intelligence branch of the Board of Education, known as the Office of Special Enquiries and Reports. In this capacity he issued a series of reports covering domestic and foreign education and did work on the 1902 Education Act.

In the beginning, Sadler and his extension team thought they were going to change the world. At the end of each course the lecturer submitted a report. In the early years we can almost see the young lecturers peering out into the audience looking for cloth caps to sustain hopes that education was being taken to the masses. At Bath, Mackinder was disappointed that none of the employees of the iron works were in attendance, but in Barnsley his hopes were raised. He commented, "In no part of England which I have worked in are our lectures more required and more appreciated than in these populous districts with their very rare opportunities for higher learning."[26] However, the extension lectures attracted audiences mainly from middle- and professional-class backgrounds. There was always a high proportion of school teachers, and the value of "teaching teachers" was not lost on the lecturers, particu-

26. Oxford University Extension Lecturers and Examiner Reports, 1886–87, vol. 2, December 22, 1887, RHO.

larly those opening up new subject areas such as geography and economics.

In the fall of 1886 Henry Bates approached Mackinder and suggested that he write down his ideas on geography in a form that might be suitable for delivery at the Royal Geographical Society. No promises were made that the paper would be accepted and no hint given that Mackinder could become part of the broader campaign, but it was widely known that the RGS was pressing the vice-chancellors of Oxford and Cambridge to establish geography teaching in their universities. Mackinder became involved in the educational debate and never returned to his legal career.

3

Royal Geographical Society

The Royal Geographical Society, London, had been founded in 1830 by explorers and travelers. By the 1880s it had a membership of three thousand fellows who were drawn from a variety of backgrounds in the professional and upper classes. Women were not eligible for election. The membership consisted of men with a general interest in the world and its affairs, officers from the army and navy, businessmen, academics, schoolteachers, diplomats, and colonial administrators. The rules concerning election to fellowships had never been tightly drawn, and over the years large numbers of those who liked to hear a good yarn of daring deeds in faraway places had been admitted. The number of fellows grew as the years passed, and the Society needed a succession of houses. Initially it occupied premises in Regent Street, then Waterloo Place (1839), Whitehall Place (1854), and Saville Row (1870); in 1913 it acquired the present house overlooking Hyde Park.[1]

In the second half of the nineteenth century the RGS was part gentleman's club and part learned society. It performed several functions, some investigatory and others social. The Society wanted to serve explorers, be a storehouse of knowledge and professional skills, promote the study of world affairs and act as a pressure group for various interests. These interests were educational, imperial, and technical where they related to the need for adequate survey and mapping.

During the last third of the nineteenth century most academic disciplines and learned societies tried to professionalize themselves. This involved defining the subject area, recognizing professional qualifications, upgrading journals, promot-

1. H. R. Mill, *The Record of the Royal Geographical Society.*

ing research and teaching, and looking more carefully at the credentials of those aspiring to membership. The Royal Geographical Society was no exception, and as might have been predicted, the attempt to professionalize geography led to strains within the membership and within the council.

As was common in many clubs there were almost two societies within the RGS: the mass of ordinary fellows and the influential group from which councilors were nominated and elected. This arrangement was institutionalized by the mechanism of allowing only the council to nominate candidates for election to the council.

When the debate started, in the 1870s, as to how the scientific standing of the Society could be improved, no one seriously considered expelling all those who paid dues but lacked scientific qualifications: the greatest part of the membership would have gone. However, there was no lack of spirited arguments on the future of geography and how the Society might be involved.

In 1876 a physical-geography proposals committee (soon to be renamed the scientific-purposes committee) was set up by the council. Francis Galton (1822–1911) got the committee to allocate money for the promotion of the "scientific branches of geography."[2] A few years later, in the early 1880s, Galton and Douglas Freshfield (1845–1934) combined to press for more educational involvement by the Society. Both men were recognized scholars and had strong connections with London's scientific establishment. Galton, in particular, was active in the affairs of many institutions, including the Royal Society, the Alpine Club, the British Association, the Royal Botanic Garden, and the Anthropological Institute.

The former naval officer and explorer Clements Markham (1830–1916) was not happy with the policies that Galton and Freshfield were advocating, as he confided on paper some years later. Markham thought that all the work of the Society was scientific and, not entirely seriously, condemned Galton's "edu-

2. Committee Minute Book 1877–83, Scientific Purposes Committee, January 3, 1878, RGS.

cational rioting."[3] Markham urged the Society to put more money into training explorers. He got his way with the appointment of an instructor of practical astronomy and surveying, but the tide of opinion was against him. In 1884 a geographical-education committee was established to promote the teaching of geography in schools, universities, and training colleges.

In the summer of 1884, Scott Keltie (1840–1927) was appointed the Royal Geographical Society inspector of geographical instruction, and with great energy he produced a survey of geography teaching in continental universities. The Society decided to publicize Keltie's findings. A hall was hired in Great Marlborough Street, a large display of teaching materials set up, and a series of lectures on geographical education delivered by James Bryce, Henry Moseley, E. G. Ravenstein, Douglas Freshfield, and Scott Keltie. The exhibition in London ran through the last weeks of 1885 and the first weeks of 1886 and was followed by displays in provincial cities. Mackinder learned of the exhibition and went to see it. By good fortune he fell into conversation with Keltie, and they had a long talk about geography and geography teaching.[4]

Keltie's report showed that the subject was extensively taught in continental universities. In Germany, for instance, there were twelve chairs of geography. To some this was a sinister development. If the continentals were better equipped to know the world, might not the knowledge be used to advance their commercial and imperial positions?

There is little doubt that the urge to promote the teaching of geography in Britain, and elsewhere, owed something to nationalist interest in territorial expansion, commercial success, and strategic questions. Certainly we can find fellows of the Society expressing views on these issues. For example, in 1886 the Marquis of Lorne, a former governor-General of Canada and author of the book *Imperial Federation*, said in his presidential address to the Society:

3. C. R. Markham, Manuscript Volume of Reminiscences, CRM 47, RGS.
4. Council Minutes, June 23, 1884, RGS; J. Scott Keltie, *Geographical Education*; M.P. Auto.

To the soldier the study [of geography] is of use, for it makes him know how to read a map, how to use it for military purposes. ... To the statesman the study of the science may mean the avoidance of many blunders, the results of which have been only too manifest in our history. Territories which are mere names to the imagination, are easily given away to become the foothold of rivals. ... The science cannot be looked at with indifference by those who direct the advance of commerce, for they in each fresh advance may find the path to new sources of wealth.

Here was the "new imperialism" ready to promote the "new geography." It would be wrong, however, to see the establishment of geography teaching as wholly an outgrowth of imperialist sentiment at the RGS. The Society was much more complex than that, as a short examination of the membership of the council will show.[5]

In the years 1884–87 fifty-one individuals served on the council of the Society. It is not possible to look into the career of every council member, but the majority were well known, and their backgrounds are recorded in standard reference works.[6] At least half of the council members in the period were, or had been, army or navy officers. More than a third had seen service in India, a fifth had held posts in the colonial or diplomatic services, and just under that fraction could be described as explorers. Many persons, of course, fell into several categories. So far it might seem that the Society's leaders fitted the caricature of the RGS as a haven for retired officers and old India men with a penchant for exploration and travel. Upon further examination, another picture emerges.

One-third of the council members were fellows of Britain's

5. Heinz Gollwitzer, *Europe in the Age of Imperialism, 1880–1914*, pp. 161–64; Marquis of Lorne, Presidential Address, *Proceedings of the Royal Geographical Society* 8 (1886): 421–22. See also Marquis of Lorne, *Imperial Federation*.
6. Information on council members compiled from standard sources, such as *Who Was Who, Dictionary of National Biography*, and RGS materials, including certificates of election; D. R. Stoddart, "The RGS and the New Geography: Changing Aims and Changing Roles in the Nineteenth Century," *Geographical Journal* 146 (1980): 190–202. Stoddart argues that in terms of scientific qualification the RGS fellowship declined during the nineteenth century.

major scientific society, the Royal Society. Of the army officers more than a third were drawn from the Royal Engineers and the Royal Artillery, which, at Chatham and Woolwich respectively, had first-class training establishments that gave a good scientific education. Of the nine Royal Engineers and Royal Artillery men who served on the council, four had been elected fellows of the Royal Society on the basis of substantial scientific investigations.

Council members representing the data-gathering side of geography, included two directors-general of the Ordnance Survey, the keeper of the printed maps and charts of the British Museum, a surveyor-general of India, the hydrographer to the Royal Navy, and the director of the Royal Botanic Garden at Kew. The garden was the center of a network of botanical stations that made Kew a major institution for the exchange of scientific knowledge. Naturally many council members had served on the governing bodies of such learned societies as the Royal Society, the Royal Statistical Society, the Royal Asiatic Society, and the Linnaean Society.

Several prominent members of the educational establishment were on the council, including the senior examiner of Her Majesty's Civil Service; the inspector-general of Military Education; Sir John Lubbock, a member of the senate of London University; and the Honorable George Brodrick, warden of Merton College, Oxford. Only about 10 percent of the council members had direct involvement in the management of banks and overseas commercial interests.

Virtually all members of the council had a wide experience of the world, and many had worked as explorers, surveyors, or scientists in distant lands. Believing that a broad understanding of the world was of educational value, they wanted to see geographical knowledge organized in such a way that it could be taught in schools and universities. Most of them would have dismissed "capes and bays" geography as failing to convey anything of the world's complexities.

Although there was no consensus within the Society as to how geography should be taught, most influential members agreed that geography should be represented in the curricula of schools and universities. At an important meeting of the

RGS council on April 12, 1886, with Lord Aberdare in the chair, measures were adopted to promote geography teaching. Oxford University Extension was given help with the expenses of Mackinder's geographical lectures. It was agreed that the vice-chancellors of Oxford and Cambridge should be contacted again to see if a place could be found for geography. In the summer of 1886 Aberdare wrote to the vice-chancellors, and influential fellows of the Society were utilized in an effort to awaken interest at Oxford and Cambridge. At Oxford, Canon Butler, James Bryce, Henry Moseley, and George Brodrick were active. In addition, J. F. Heyes, who had won one of the Society essay prizes when a schoolboy, wrote an article supporting geography in the *Oxford Magazine*.[7] At Cambridge another prize winner, Donald MacAlister (1854–1934), promoted the subject. By the end of 1886 both Oxford and Cambridge were interested in what the Royal Geographical Society was saying about the need for geography in the curriculum, especially because the Society wanted to fund new teaching positions.

The Society needed scholars capable of teaching geography as a broad, synthesizing subject at the university level. In the history of an idea there comes a point at which many can sense the next development but only a few are capable of operationalizing the new thoughts. In this context, Mackinder, the young lawyer who lectured on physical geography for Oxford University Extension and had training in natural sciences, history, geology, economics, and international law, was a godsend. Elected to the Society in March 1886, Mackinder lunched occasionally with Bates, Keltie, and Freshfield. There can be little doubt that talks with these officers of the Society helped form Mackinder's ideas on the possibility of creating a unified subject that incorporated both physical and human geography. It was Henry Bates, famed for his natural history work in Amazonia and his scientific association with Alfred Wallace, who

7. J. R. Heyes, "A Plea for Geography," *The Oxford Magazine*, December 8, 1886. Some of the phrases and ideas used by Heyes are similar to those used in "Scope and Methods" and by Markham in his report on the paper. Perhaps in orchestrating the campaign for geography, senior members of the RGS showed Heyes the Mackinder paper in manuscript form.

suggested to Mackinder that he write down his ideas on the nature of geography.

In November 1886 the Royal Geographical Society received Mackinder's handwritten paper "On the Scope and Methods of Geography."[8] The paper defined geography and showed how the subject could be developed in the educational system, and it brought the whole debate, on where geography and the Society were going, to a head. But it was more than a polemical piece, for it showed by example how geography could be made a bridging subject in the educational system. The scientific group on the council were so pleased that they seem to have had no reservations when the paper was assigned to Clements Markham for review at the meeting of November 22, 1886.

Markham's report on the "Scope and Methods" paper was crucial, for he had doubts about the Society's getting involved in what he regarded as educational sideshows. Markham did not let his view of policies he distrusted interfere with his evaluation of the paper. For Markham the paper was, in some respects, the most important that had been communicated to the Society during the twenty-five years he had been on the council. Never before had the scope and objectives of geography been so lucidly defined. Markham observed:

> The question which Mr. Mackinder discusses is whether the science of geography is one investigation; or whether physical and political [human] geography are separate subjects to be studied by different methods, the one an appendix of geology, the other of history. He contends for the former view and that no rational political geography can exist which is not built upon and subsequent to physical geography. . . .
>
> One of the greatest gaps lies between the natural sciences and the study of humanity. It is the duty of the geographer to build a bridge over this abyss, which is upsetting the equilibrium of our culture.[9]

8. H. J. Mackinder manuscript, "On the Scope and Methods of Geography," RGS. Markham's report on the paper is attached.
9. "Clement Markham's Report on a Paper by Halford Mackinder," *Geographical Journal* 147 (1981): 269–71.

The idea that geography might serve as a bridge between the sciences and the humanities was in the educational air of the 1880s, and Heyes referred to it in the *Oxford Magazine* in December 1886. Nevertheless, Markham's evaluation was shrewd, and he concluded it by recommending that the paper be presented at a meeting of the Society, published in the *Proceedings*, and discussed at a special meeting.

The paper was delivered on Monday, January 31, 1887. "What is geography?" was the opening line, and it caught the attention of the audience, including the army officers and sea captains who did not see the need for the question, particularly when it came from a young man who lacked experience overseas. Old Admiral Erasmus Ommanney (1814–1904), who had taken part in the battle of Navarino (1827) as a teenager, sat in the front row muttering "damn cheek, damn cheek." But Union debaters enjoyed stirring up the audience and Mackinder was not put off. On he went and attacked a prominent council member, Sir Frederic Goldsmid (1818–1908), for his statements at the British Association, which had suggested that scientific geography would have to mean physical geography and that human geography, with its strong historical element, could not be incorporated. To Mackinder the subject could not exist unless it had human and physical elements.[10]

The strongest section of the paper was the attempt to construct a geography of southeastern England. Combining his experience of the area as a schoolboy, and as a member of the Rifle Volunteers, and drawing on wide reading in geology, Mackinder outlined the geomorphology of the area, and proceeded to show how the anatomy of the land had been an influence on human activities. He was careful to avoid the idea that the physical background had determined a human response, point-

10. H. J. Mackinder, "On the Scope and Methods of Geography," first published in the *Cambridge Review* 8, no. 196 (March 2, 1887): 247–49, and no. 197 (March 9, 1887): 264–67. MacAlister requested permission to publish as part of the effort to raise interest in geography at Cambridge (Later the paper was published in the *Proceedings of the Royal Geographical Society* 9 [1887]: 141–60. Discussion of the paper covers pp. 160–74.); F. J. Goldsmid, Presidential Address, Section E, *Report of the Fifty-Sixth Meeting of the British Association held at Birmingham, September, 1886*, pp. 721–26.

ing out that the importance of geographical features varied with the degree of civilization and with technological changes. What was more, man altered his environment and "the action of that environment on his posterity is changed in consequence." The course of human history was a product of environment and "the momentum acquired in the past."[11]

After moving to a different scale, and showing how the geography of the Indian subcontinent could be developed, Mackinder concluded by stating, "I have sketched a geography . . . which will satisfy at once the practical requirements of the statesman and the merchant, the theoretical requirements of the historian and the scientist, and the intellectual requirements of the teacher."[12] Where had Mackinder's ideas on geography come from? They had emerged via Henry Moseley (and to a lesser extent Bates and Galton) from a British scientific-exploratory tradition that had striven to comprehend the physical environment of the globe as a whole. One aim of the tradition had been to develop knowledge of the physical environment as a means to better understand the distribution, and evolution, of plants and animals. As Charles Darwin had told the botanist Joseph Hooker in 1845, geographical distribution was the grand subject that was almost the "keystone of the laws of creation." When Mackinder defined geography, in the discussion of his "Scope and Methods" paper, as "the science of distributions," he was drawing on a biological tradition of which Moseley was a part.

Onto the biological tradition Mackinder grafted a social-

---

11. "Scope and Methods," p. 157. Sometimes, on the basis of his statement that human geography must be "built upon and subsequent to physical geography," Mackinder is labeled an environmental determinist. A recent instance is to be found in Arild Holt-Jensen, *Geography: Its History and Concepts*, p. 26. This evaluation ignores one of the major arguments of Mackinder's paper, that geography is a subject with human and physical dimensions and unless one recognizes that, the subject will indeed be divided between history and geology. In the section of the paper quoted by Holt-Jensen, Mackinder is trying to convince skeptics in his audience that a unified subject exists, not that the physical environment determines human activity, a notion Mackinder rejects later in the paper.
12. "Scope and Methods," p. 159.

science element at just the time when such subjects were coming into university curricula. Mackinder's geography was based on an evolutionary, worldwide view, in which forces interconnected and played upon each other. Although the phrase was not in use, he was employing an ecosystem approach.

Mackinder's ideas also grew out of his personal experiences: the boyhood in Lincolnshire and Surrey and his father's interest in medical geography. Ideas implanted by experience flourished when they came into contact with the intellectual traditions of the time during his five years in Oxford. Mackinder's ideas were not transplanted from Germany, nor from any other European country that had chairs of geography. The works referred to in the "Scope and Methods" paper were largely British, and the German geographers were dismissed with the remark that they included "too much in geography." Peschel's *Physische Erdkunde* was referred to merely to illustrate that point.[13] Later he came to know German geography and geographers well.

Were Mackinder's ideas on teaching geography new, or was he setting up traditional geography, with its recitations of ports, harbors, rivers, and towns, as a paper tiger that he could ignite amidst an exciting lecture performance? In the 1880s physical geography was already well taught in some quarters. Mary Somerville's *Physical Geography* had been available since 1848 and had gone into several editions. Thomas Huxley's *Physiography* had been published in 1877[14] and was used extensively in conjunction with the science teaching promoted by the Department of Science and Art in South Kensington. Mackinder's teaching in physical geography reflected this material, but he was ahead of his time in his ability to see dynamic interconnections between forces in the physical environment.

In human geography (or political geography as it was referred to in the 1880s) there was only a gazetteer approach to the subject. Gazetteer geography concerned itself with locating places and knowing what was produced there. Archibald

13. Ibid., p. 154.
14. D. R. Stoddart, "The Victorian Science—Huxley's 'Physiography' and Its Impact on Geography," *Institute of British Geographers, Transactions* 66 (1975): 17–40.

Geikie's *Geography of the British Isles* (1888) illustrated the static approach well. There was a description of physical features, which relied heavily upon geology, and then a county-by-county account of human activity. For example, Oxfordshire for Geikie was "a long strip of country on the north side or left bank of the Thames. . . . It is essentially an agricultural county. Its chief town, Oxford, is the seat of a bishopric and of a famous university. Other towns are Banbury, long noted for its cheese and cakes; . . . Witney, famed for blankets."[15]

Mackinder's paper did not stop this approach. (Geikie's book was printed six times between 1888 and 1907.) Not until the appearance of *Britain and the British Seas* (1902) was a new method fully displayed, and it was 1906, with the appearance of *Our Own Islands*, before Mackinder's method had an impact at the school level.

At a key point, when groups in Britain, partly for trade and imperial purposes, wanted to know more of the world scene, Mackinder pulled together recent developments in the natural and social sciences to create an approach to geography that examined man-culture-environment relationships in a broad way. Mackinder had the intellectual training to make the bridge between the natural and the social sciences. But knowledge was not enough. Mackinder's strongest mental attributes were vision and imagination, and they allowed him to forge a new discipline in English higher education.

Two weeks after the delivery of the paper, a special session was held to discuss it. In a conversation with Mackinder before the event, Francis Galton suggested that the young man be less rhetorical and a bit more humble. Tactics were chang-

15. Archibald Geikie, *An Elementary Geography of the British Isles.* Stoddart, "RGS and New Geography," discusses the paradox of Geikie's noninvolvement with the RGS. The simple answer seems to be that he believed that geology obviated the need for geography as a separate discipline. In lectures in Edinburgh he told his students: "Our consideration of Physical Geography will be more particularly in regard to its geological bearings. A great deal of matter in Physical Geography textbooks is simple fanciful speculation without philosophical foundation" (Lectures Taken Down by Ralph Richardson 1871–72, Gen. 694, Sixth Floor Collection, Edinburgh University Library).

ing; now was the time to let the other side have a say and evolve a common policy. On the evening of February 14, 1887, Sir Frederic Goldsmid defended his ideas and was graciously received. Most of the remaining speakers, several specially invited, supported Mackinder's views. T. W. Dunn, headmaster of Bath College, whom Mackinder knew well as a result of his extension work in the west, was a strong supporter, as were Francis Galton, James Bryce, and H. G. Seeley. Moseley could not be present because he had commitments in Oxford. The last major speech came from Douglas Freshfield, who read at length from Markham's report on the paper, which helped bring the factions together again. Nevertheless, it was no doubt helpful that Markham was away in the West Indies at the time.

Spectators of these events had enjoyed some fine entertainment. In the view of *The Times* the Society had been challenged to decide if it wished to be a scientific society or a club for those interested in travelers' tales. Mackinder's presentation had shown that geography could be treated scientifically, and the paper had "received the substantial approval of all the most influential members of the society."[16] Whether the tone of the article owed anything to the journalist connections of members of the reform group, such as Keltie or Brodrick, is not known, but if reactionary elements wanted to fight, they would clearly label themselves, in the opinion of *The Times*, as not being among "the most influential members of the society."

Events now moved forward rapidly. On February 28, 1887, the council was told that the delegates of the Common University Fund, at Oxford, had agreed that a reader in geography should be appointed, for a five-year term, at an annual salary of £300. On May 9 the council agreed to pay half the salary for five years, provided it had a voice in the appointment, and the university accepted this stipulation. At the June 6 council meeting it was announced that the electors appointed by the university were W. Ince (Christ Church), Brodrick, Bryce, Moseley, and V. Harcourt (Christ Church). The Society was to be represented by Francis Galton and W. T. Blandford (1832–

16. *The Times*, February 18, 1887.

1905). Overwhelmingly the selection committee knew Mackinder. There were other candidates for the readership, but Henry Moseley took a lead and insisted upon a man with a strong background in natural sciences and physical geography. There was at this time, in the Society and the university, a feeling that the geography taught should be scientifically based, and there were hopes that physical geography could be brought into the examinations for degrees in natural sciences. On June 27, 1887, the council of the Royal Geographical Society learned that Halford J. Mackinder was to be appointed to the Oxford readership in geography.

Events at Cambridge took a similar course, but it was difficult to find candidates for the lectureship.[17] Mackinder was approached and was prepared to move to Cambridge if he did not win the Oxford post.[18] One of the key figures in the establishment of geography at Cambridge was the professor of zoology, Alfred Newton (1829–1907). The first appointment went to F. H. H. Gillemard (1852–1933), an explorer and naturalist with a medical training. He resigned after a few months, and J. Y. Buchanan (1844–1925), an oceanographer who had been with Moseley on *Challenger*, was appointed.

The RGS gave Mackinder a £50 grant for travel to Germany in preparation for his work at Oxford.[19] By September he was back in Britain, at the Section E meeting of the British Association in Manchester, where Professor W. Boyd Dawkins delivered a paper on "The Beginnings of the Geography of Great Britain." This was a study in historical geology, and Mackinder made the point, quite sharply, that most of what Dawkins had said was irrelevant to geography.[20]

A constant problem in the early days of British academic geography was to prevent the dynamic discipline (that the biological-exploratory tradition had created) from being taken over by geology and, to a lesser degree, history and turned into

17. D. R. Stoddart, "The RGS and the Foundation of Geography at Cambridge," *Geographical Journal* 141 (1978): 216–39.
18. Committee for University Extension June 6, 1885–Dec. 5, 1889, March 3, 1887, RHO.
19. Council Minutes, June 27, 1887, RGS.
20. *Manchester Guardian*, September 6, 1887.

a static, deterministic subject. The geologists were liable, when they attempted geographical explanations, to be simplistic. Economic activity could be understood from the geology map. Rich farming areas were on lands with good soils, which in turn depended upon underlying geology. Industry was where coalfields were. And so on. Historians, who dabbled in geography, were prone to deterministic utterances. H. B. George in his *Relations of History and Geography* (1901) was still claiming that "no one will deny, however firmly he insists on believing in free will, that the destinies of men are largely determined by their environment . . . at any rate before extensive commerce has been developed. . . . The physical features of the earth, sea, mountains, etc., go far to fix their occupations."[21] Although British geography strayed toward this type of thinking after 1892 and the publication of Friedrich Ratzel's *Anthropogeographie,* it is not present in Mackinder's "Scope and Methods" paper. In fact Charles Darwin, *The Origin of Species* (1859), had undermined the determinists' position before they occupied it. Darwin wrote, "There is hardly a climate or condition in the Old World which cannot be paralleled in the New. . . . Not withstanding this general parallelism in the conditions of the Old and New Worlds. How widely different are their living productions!"[22] It is difficult to overstress the importance of biology on the emergence of British geography. For reasons that are not clear, the influence of the nineteenth-century naturalists is only dimly recognized. This is surprising, because many postwar developments in geography have strong biological antecedents: systems approach, spatial analysis, and the study of ecosystems.

At the British Association meeting in Manchester, in 1887, Mackinder outlined his ideas on the geography curriculum he intended to establish at Oxford. His presentation gave Dawkins and Seeley the chance to say that he was not including enough geology. The meeting offered him opportunities to hear papers by people who wanted to promote the teaching of geography in schools. (Out of this group the Geographical Asso-

21. H. B. George, *Relations of History and Geography,* p. 7.
22. Charles Darwin, *The Origin of Species,* p. 338.

ciation was to emerge.) In October 1887 he began lecturing to undergraduates at Oxford.

At Manchester, Mackinder met the Reverend David Ginsburg, LL.D. (1831–1914), the renowned Old Testament scholar. Halford was invited to Virginia Water, Surrey, where the Ginsburgs lived, and he began to pay court to one of the four daughters in the household: Emilie Catherine, who was known to everyone as Bonnie. The courtship flourished. In June 1888 the couple were engaged and a wedding arranged for January 4, 1889.

The wedding, a rather grand affair, was reported in a number of newspapers, including the *Daily News*. There were special saloon railway carriages for guests traveling from London, the station at Virginia Water was decorated, and the church still had its Christmas finery. The service was performed by a troop of clergymen that included the Bishop of Winchester and the Dean of Peterborough. The newspapers wrote the event up in a florid style with lists of guests and gifts and with details on the dress and flowers of the ladies. One local newspaper, in its efforts to please, ascribed considerable powers to Ginsburg, indicating that he was "one of the revisers of the bible."[23]

After the ceremony the couple traveled to Menton, in the south of France, where Mackinder's former headmaster, the Reverend William de Lancy West, was a chaplain. The holiday was fairly short; Halford had to be back in Oxford for the start of the Lent term. The couple moved into Bradmore Road, just north of the University Parks. To all appearances they were set for the busy, pleasant life of a north Oxford academic family. It was not to be. The marriage, which never would become a lasting alliance, soon was scarred by tragedy. In the spring of 1890 Bonnie conceived a child, and on New Year's Day, 1891, a son was born. The boy lived for eleven hours. The next day Mackinder registered the death of the unnamed and unchristened child.[24]

23. *West Middlesex Times*, January 5, 1889; *Daily News*, January 8, 1889.
24. Register of Deaths, 3a, 541, 1891, General Register Office, St. Catherines House, London.

4

Founding Reading University

After Halford Mackinder's appointment to the Oxford reader-ship in geography in 1887, he had two major educational com-mitments. First, he developed geography teaching at the uni-versity, which culminated in the opening of the School of Geography in 1899. Second, he continued with extension work, for Michael Sadler, and this resulted in the creation of Read-ing University.

Michael Sadler took over Oxford University Extension in 1885 and expanded the operation. By 1890 he had succeeded in appointing Hudson Shaw and Mackinder as staff lecturers. However, it was obvious that a program that sent academics to give occasional lecture series in towns was not going to have a widespread impact on educational standards. Sadler saw the need to create a more substantial presence in suit-able towns, and out of this grew the idea of an extension col-lege at Reading.

For Mackinder, 1892 was to be a critical year. The five-year support provided by the Royal Geographical Society was to be reviewed and, it was hoped, renewed by the council. At Read-ing the extension-college idea would either crystalize or crum-ble. In the spring of 1892 he was asked if he wanted to move to a university in the United States.

The American Association for the Extension of Univer-sity Teaching held its first annual meeting in Philadelphia in 1891. Michael Sadler attended and delivered a paper. In 1892 Mackinder was invited. He made an elaborate trip, without Bonnie, which involved teaching in the University of Pennsyl-vania summer-extension program for six weeks, attending the extension conference, and seeing something of the East and Midwest.

Mackinder arrived in New York on the *Teutonic* on March 9, 1892, and traveled to Philadelphia. His reputation as an exciting lecturer preceded him, and both the *Philadelphia Inquirer* and the *Public Ledger* ran stories. The *Ledger* carried a picture of the young lecturer who had "practically revolutionized the popular teaching of geography in England."[1]

While in Philadelphia, Mackinder gave six lectures on "Revolutions in Commerce." Each week he delivered a new lecture and made a presentation at three different locations. Toward the end of one series he started another, "The Unity of History," for the University Lecture Association. In addition he gave occasional lectures at Swarthmore College and at the Drexel Institute. The Drexel lecture, entitled "The Teaching of Geography," was aimed at schoolteachers. The presentation did not receive long reports in the newspapers, but the *Public Ledger* did record Mackinder's definition of the subject: "Geography . . . was a definite and graphic science, the object of which was to study space relations on the earth's surface, together with their cause and effect." It may be a mistake to assume that Mackinder's view of geography was quickly lost in the newspaper files of Philadelphia. The whole question of the place of geography in the American school curriculum was examined in 1892 as part of the activity of a Committee of Ten appointed by the National Educational Association. The committee was organized into nine conferences, one of which considered geography. The conference on geography included Edwin J. Houston, of Philadelphia, who probably had the report of Mackinder's lecture brought to his notice, if he was not present at the delivery.[2]

After completing his work in Philadelphia, Mackinder went to see how geography was taught at Harvard, Princeton, and Johns Hopkins universities. He found the Harvard laboratories, under W. M. Davis, especially striking and met some of the students that Davis was training. Davis and several of his colleagues later taught in Oxford summer schools, and the visit

1. *Public Ledger*, March 9, 1892.
2. H. J. Mackinder, *Revolutions in Commerce; Public Ledger*, April 7, 1892.

helped Mackinder define his ideas about the type of school of geography he would like to establish in England.[3]

The greatest part of Mackinder's time in the United States was spent in the East, but he made a journey into the Midwest to look over the possibilities of a position at the University of Chicago. At that time the first president, William Rainey Harper, was recruiting staff in readiness to admit students in the fall of 1892. With a generous endowment the new university was offering high salaries to induce scholars "to go off into the wilds" as one Yale man exclaimed.[4]

It is not certain how Mackinder and Harper made contact. Harper had visited Oxford in 1891, and he was in Philadelphia in the spring of 1892 lecturing on the first twelve chapters of Genesis and trying to persuade Professor E. J. James, of the Wharton School of Finance, to join the University of Chicago. Harper wanted a strong extension program and had persuaded R. G. Moulton, of Cambridge University Extension, to join Chicago. Moulton acted as go-between, putting prominent members of the British extension movement in touch with Harper. In any event Mackinder had made contact with the University of Chicago, and after working in the East he set off for the Midwest.

As he traveled westward, Mackinder stopped briefly in Washington and dispatched a telegram to Reading in reply to the suggestion that he head the new extension college there. The message simply read, "Yours received important postpone decision home May eleventh, Mackinder."[5] He did not feel it necessary to add that in the meantime he was going to Chicago to talk with the president of another new institution about a post.

From Washington, Mackinder crossed the Allegheny Mountains and went north across Ohio to Toledo, where he delivered a lecture, "The Origins of Civilization," and spoke briefly

3. Mackinder to RGS, May 21, 1892, University Archives, SG/R/I/I, Bodleian Library, Oxford (Bodleian).
4. James Dana to W. R. Harper, July 13, 1890, W. R. Harper papers, University of Chicago Archives (UCA).
5. Mackinder to Palmer, April 20, 1892, UCR 8, Reading University Archives (RUA).

of the university-extension movement in Britain. From Toledo, Mackinder journeyed westward to Chicago, where, in the sparse words of his own account, "President Harper . . . tried to persuade me to throw in my lot with him but I am glad to say in vain." There is little detail about the Chicago trip other than mention of visits to the grounds where the Columbian Exposition was to be held, in 1893, and to the Wheat Pit.[6]

After Chicago, Mackinder went by train to Montreal and then, via Niagara, back to New York for the return voyage. It is intriguing to speculate what would have happened to American geography had Mackinder joined Chicago. The department of geography that did emerge there was based heavily on geology. Mackinder, with his emphasis on the biological sciences and history, could have been an important broadening influence. Would he have gone to Paris in 1918 with the American geographers Isaiah Bowman, Mark Jefferson and Douglas Johnson? The speculations are endless once President Wilson's democratic ideals are placed along side Mackinder's grasp of European realities.

The Extension College at Reading

Michael Sadler wanted to set up university extension centers in several towns, but this required greater financial resources than he commanded. In the summer of 1890 an opportunity emerged to deal with the money problem.

The local taxation (custom and excise) act of 1890 included a provision that the duty raised on spirits—the "Whiskey money"—should be used to provide county councils with funds to support technical education. Mackinder and Sadler read this in their morning papers, and both had similar thoughts: Perhaps some of the money could be diverted to university

6. *Toledo Daily Blade,* April 25, 1892; *The Bee,* April 23, 1892; M.P. Auto. There are no letters from Mackinder to Harper in the University of Chicago archive. When Harper spoke to Mackinder, T. C. Chamberlain and R. D. Salisbury were already under consideration at Chicago. An extensive Geddes correspondence exists at Chicago. Geddes offered the University of Chicago a College of Scotsmen, but this proved a bit too much even for Harper's imperial instincts (Geddes to Harper, November 16, 1891, and February 20, 1892, UCA).

extension teaching and used to build extension centers in suitable towns. Mackinder recalled the July morning vividly, when Sadler and he laid their plans:

> Each set out to consult the other and we met half way. I have a clear picture in my memory—it was beside Keble College that we met. A few words, for we well understood each other, and we decided to issue a book. There and then we went off to the Clarendon Press and asked the somewhat astonished secretary if he would publish a book for us.
> Yes, he replied.
> In a week?
> Is it written?
> No.
> None the less we promised the manuscript and he promised to print, and on both sides we were as good as our word.
> Sadler and I worked together in an upper room in Sadler's house through all the waking hours of several days. Then it was that the idea of a "University Extension College" first began to shape in our minds.[7]

The result of these efforts was a short work entitled *University Extension: Has It a Future?* published in the autumn of 1890. The little book was quickly sold out, reprinted, and in 1891, published in revised form. The first edition argued for the creation of extension centers close to public libraries in towns that lacked a university or college. In the second edition the idea of creating a University Extension College was fully worked out, to the point of perspective drawings and the floor plans of proposed buildings.[8]

The notion of creating academic communities away from the university was not new. Toynbee Hall, set up in 1884 in the east end of London, had strong links with Balliol College. Benjamin Jowett (1817–93), master of Balliol, had suggested

7. Notes on the Origin and Early Development of University College Reading, Now the University of Reading, March 28, 1926, Mackinder Papers, School of Geography, Oxford.
8. H. J. Mackinder and M. E. Sadler, *University Extension: Has It a Future?* The book later was revised as *University Extension, Past, Present, and Future.*

for years that the work of the university should be more wide-spread. He talked of making choices between education and revolution and spoke of sending out scholar missionaries who would eventually settle in towns where their work was needed. Jowett was a supporter of extension work and the creation of provincial universities. He encouraged innovative young men to develop new educational ideas. Several of the extension lecturers engaged by Sadler were Balliol men, and when he organized the first extension summer school, in 1888, Balliol provided facilities and Jowett acted as a father figure to the students who came to the university for a few weeks.

Not everyone in Oxford approved of the missionary work. Older dons complained about the loss of summer quiet and, in general, found the Balliol causes too progressive. Oxford is a conservative place. Many younger scholars were against change, too. For example, the historian Charles Oman (1860–1946), who was in the same age group as Mackinder and Sadler, was unenthusiastic about the efforts of Jowett to link the work of the univeristy to the broader needs of late nineteenth-century Britain.[9]

Colleges other than Balliol also were prepared to encourage experimentation. In 1886, for example, Michael Sadler was appointed steward of Christ Church with the understanding that he would devote a portion of his time to extension work. Late in 1890, with members of the senior common room, he began to explore the possibility of establishing an extension college in a suitable town. It was not long before the name of Mackinder, a recent undergraduate at Christ Church and now an established extension lecturer, was suggested as head

9. Charles Oman, *Memories of Victorian Oxford*, p. 234; Notes of Conversation with Sadler on Thursday, October 31, 1889, Hewins Papers, 43/154, University of Sheffield. There were rivalries within Sadler's university extension group. Hewins, appointed in 1888 to run the extension summer schools, was junior to Hudson Shaw, Mackinder, and Sadler. In attempting to advance his standing, he irritated Sadler. Hewins argued with Sadler about these issues in the fall of 1889 and recorded his thoughts on paper. At the time Hewins tried to ease Mackinder out of the summer school organization but found Sadler loyal to Halford. Hewins recorded that Mackinder was not popular in Oxford. This probably was true—Mackinder had collected rather too many degrees, qualifications, and awards by Oxford standards.

of the new college. That he already held the readership in geography at the university was important, for it meant that the extension experiment could be started by simply finding a small additional income for Mackinder. The location would have to be close to Oxford if Mackinder were to run the college and be reader in geography.

In extension circles there was discussion as to the best town to start the experiment. Banbury had technical-education facilities but was rather small. Leamington Spa was too close to the emerging university of Birmingham. Swindon and Newbury had advantages but could not be reached directly from Oxford by train. Oxford extension lecturers attracted huge audiences in the West Riding of Yorkshire and in such southern resort towns as Brighton, Cheltenham, and Bath. All, however, were too far away if the principal had to lecture in Oxford; furthermore, the resort towns lacked the working-class audiences that Sadler and Mackinder wanted to reach. The extension program at Reading had been limited, but Sadler had reactivated work there in 1887. The main advantage of the town as a location for an extension college was that it was a half-hour by train from Oxford, and Oxford scholars could be used to strengthen a teaching program.

Had the location been selected without reference to Mackinder's duties in Oxford, Bath would have been the place. Several extension lecturers had large followings there, and links had been established with T. W. Dunn's school, Bath College. Officials in Bath wanted a college and did establish the Merchant Venturer's Technical College in 1894. The college would be chartered as the University of Bath in 1966.

Reading was not served by any other university or college, nor was it likely to be. It was the county seat of Berkshire and was close to the counties of Hampshire, Buckinghamshire, Surrey, Wiltshire, and Oxfordshire. Mackinder and Sadler suspected that, like many county councils newly established by act of Parliament in 1888, the Berkshire council would not be in a position to offer the technical education provided for in the local taxation act of 1890 and therefore might be interested in a cooperative venture. Furthermore, Reading had a small School of Science and Art that prepared students for examinations set by the government's department of science and art in South Kensington, London. The school would be an asset in any effort to capture technical-education funds.

In the fall of 1890 Sadler contacted Walter Palmer (1858–1910) of the university extension association in Reading. Palmer was a member of the family that had built up substantial biscuit factories in Reading, and by the spring of 1891 Sadler had allied him to the idea of creating an extension college. Scientific and literary societies in Reading also expressed interest. To bring everyone together Walter Palmer held a *conversazione,* at the town hall, on October 21, 1891. Sadler asked Max Müller and Mackinder to attend.[10]

The *conversazione* was a lengthy affair. The Berkshire Amateur Orchestral Society plowed through a long program. The interval was devoted to the main business of the evening. Max Müller spoke first and in an easy way outlined the idea of establishing a small college as an offshoot of the great university in Oxford. Then Mackinder, in a brisk manner, explained how the extension college would operate. Max Müller and Mackinder, who knew each other well, provided the right mix of the great scholar floating out fine thoughts, and the enthusiastic young man showing that the proposed college would be tightly organized and efficiently administered.

The idea of an extension college took root quickly, but there were numerous parties to be satisfied: Oxford University Extension, the Reading Extension Association, Christ Church, the local school of science and art, and the Reading Town Council. In the spring of 1892, Christ Church elected Mackinder to a studentship (fellowship), and Francis Paget, dean of Christ Church, wrote to Palmer, on May 21, 1892, offering Mackinder's services to the new college.[11] Paget's letter was presented to the annual meeting of the Reading Extension Association. Mackinder attended the meeting and told the group that it

10. Sadler to Walter Palmer, November 11, 1890, UCR 2, RUA; *Berkshire Chronicle,* October 24, 1891.
11. Paget to Walter Palmer, May 21, 1892, RUA. This letter is the founding document for the University of Reading. In the letter Paget refers to Reading as the "oldest centre of University Extension in the Oxford district." Tactfully he did not mention that Sadler

was entirely up to them whether or not they accepted the Christ Church offer and took him on as principal. He would not cost anything in salary, and the fees he earned from teaching courses in Reading could help to employ additional lecturers. Mackinder concluded that the new college "must meet the wants of all classes" and increase opportunities for the "so-called masses." The association accepted the Christ Church offer and, at a subsequent meeting, terminated itself and handed over a modest bank balance to the new college. Several prominent members of the association, including Walter Palmer, joined the council of the extension college.[12]

The Reading Town Council and the committee that ran the School of Science and Art agreed, on June 2, 1892, to amalgamate the school with the new extension college.[13] The school, which had premises, students, and a few part-time staff members, qualified the extension college for technical-education funds. Two key individuals came with the school. Francis Henry Wright, who had been secretary to the school, set in place the administrative structure of the new college and eventually became registrar of Reading University. Herbert Sutton had been vice-chairman of the committee that ran the school. He became chairman of the extension college council and was to lend from his own fortune the money that allowed the college to become independent of the town council and a university rather than a technical college.

The extension college inherited the premises of the School of Science and Art. The major building was the former hospitium of the Order of St. John, which stood in the center of

had reactivated it in 1887. There was a clever effort on the part of Paget, Sadler, and Mackinder to give the impression that the extension college was a natural outgrowth of existing extension work and the Reading School of Science and Art.

12. *Berkshire Chronicle*, June 4, 1892.
13. Mackinder met with the council of the School of Science and Art on June 2, 1892 (*Minutes of Council*). At the meeting he was appointed Director of the Reading School of Science. This development gave Mackinder day-to-day control of the small physical and teaching resources of the School of Science and Art. Mackinder's first-class degree in natural science was a strong qualification for the position. There was an intense period of negotiation between various parties in Reading in June 1892.

the town and overlooked the churchyard of St. Laurence. In addition some rooms in the nearby municipal building were used for art classes.

Opening the Extension College

On Michaelmas Day, 1892, Francis Paget, dean of Christ Church, opened the University Extension College in Conjunction with the Schools of Science and Arts, Reading, with Mackinder as the first principal. The college started with fewer than twenty staff members and a few hundred students. Everyone worked on a part-time basis. Courses were offered in history, geography, biology, chemistry, physics, mathematics, music, art, shorthand, and typing. The staff of the School of Science and Art numbered seven, and most stayed on, including the biologist B. J. Austin (FLS), who had been appointed in 1871. He became the first librarian of the college in addition to teaching biology. New part-timers joined the staff. G. J. Burch, who had taught science and physiography for Oxford University Extension, came to teach physics and chemistry.[14] Later he became a fellow of the Royal Society and was made Reading's first professor of physics, in 1907, when a professoriate was created. Some well-qualified scholars who lived in Reading were drafted onto the staff. J. C. B. Tirbutt, who

14. *Berkshire Chronicle,* October 1, 1892; University Extension College, Reading, Calendar, 1892. As with Oxford University Extension under Sadler, the records of the Extension College at Reading under Mackinder are well kept. They are housed at Reading University. Pertinent records include: University Extension Association Minutes, 1891–92; Minutes of Council, 1891–95 (the early minutes cover the work of the School of Science and Art); Notes on Donations and Expenditure, 1892–1901; List of teaching staff and qualifications, 1895; University Extension College Journal, 1894–98; Reading College Journal, 1898–1900; Court of Governors, 1896–1926; Council Minutes 1, 1896–1901; Council Minutes 2, 1901–1905; University Extension College Analysis of Expenditure, 1896–1904, Analysis of Receipts, 1896–1900 (University Extension College, Reading College, University College Reading, University of Reading: Some Early Letters 1889–1927). This valuable annotated list covers much of the important early correspondence. I am greatly indebted to J. Edwards and Michael Bott, of Reading University, for unfailing assistance on visits to the archive.

had a music degree and was active in Berkshire music circles, taught music theory and established the school of music. Catherine Pollard (1869–1960) was living with her parents in Reading after a brilliant undergraduate career in natural sciences at Oxford. She helped Austin with the teaching of biology.[15]

From this modest beginning the college grew rapidly. By 1895 there were thirty staff members, still largely part-time but offering many more courses. A. H. Green and H. W. Dickson traveled from Oxford to lecture on geology and meteorology. D. A. Gilchrist and H. P. Foulkes, both with degrees from Edinburgh, lectured on agriculture; shortly, Frank Walker, from the same university, joined them. Walker persuaded a family friend, Frank Stenton, to come to the extension college to prepare for university entrance.[16] In 1898 there were forty-four staff members, of whom twelve were on full-time appointments.

The most rapid growth was in agriculture and related subjects. In 1893–94 a department of agriculture was organized, and dairying work was offered. Surrounding county councils provided financial aid. In addition the Board of Agriculture and Fisheries became an ally. The permanent secretary to the board was Thomas Elliott (1855–1926). Mackinder and Elliott, who first met at the British Association meeting in 1889, had become friends. As the program at Reading grew, Mackinder told Elliott it was a scandal that while colleges in the marginal lands, such as Bangor and Leeds, were given money, the really agricultural parts of Britain did not share in the board's grants. The board responded promptly. In 1893–94 the college received £150, which was raised to £500 in 1894–95 and subsequently increased to £800. On May 22, 1894, Convocation of the University of Oxford approved the establishment of a committee drawn from University Extension and the college to organize instruction and examination in agriculture. In 1896 Convocation approved the award of a diploma in agriculture at Reading.

15. Chalmers Mitchell to C. Pollard, June 13, 1891, Childs Papers, RUA.
16. Doris H. Stenton, "Frank Merry Stenton 1880–1967," *Proceedings of the British Academy* 54 (1968): 315.

In 1895 the British Dairy Farmers Association decided to move the Dairy Institute to Reading, and the college opened a new building for the purpose in 1896.[17] In the academic year 1897–98, agricultural extension work was started in the surrounding counties. A horticulture department was organized in 1902, and in the following year a farm was purchased at Shinfield for agricultural experimentation. When Mackinder looked back to the early years of Reading, he thought that the success in attracting funds to develop agricultural subjects had given the college a viable financial base.

Growth in agriculture did not mean that purely academic subjects were neglected. French and Italian were added to the curriculum in 1893 when Monsieur J. Maurice Rey joined the staff. In 1895 the literature department was strengthened and commercial classes added to the existing work in shorthand and typing. The following year William George de Burgh (1866–1943) became lecturer in classics. In 1897 the school of music was established, the fine art department was enlarged, and de Burgh added philosophy to the curriculum.

No doubt Mackinder and Sadler did not speak too loudly, in some parts of Oxford, about the work in agriculture and dairying. They were pioneering a type of institution of higher education akin to an American state university, or land-grant college. Sadler in particular admired American models of education, which he felt were designed for the needs of a modern society. The United States was visited by Sadler in 1891, by Mackinder in 1892, and by Hudson Shaw in 1893 and 1894. Today the University of Reading, on the Whiteknights campus, acquired only in 1947, has the spacious qualities of a midwestern university. Even the surviving Waterhouse buildings are reminiscent of domestic architecture in nineteenth-century midwestern towns.

Many of the present-day academic characteristics of the university are a product of the period 1892–1903. The strength of Reading in agriculture, food science, agricultural extension, rural development, agricultural economics, dairying, botany,

---

17. British Dairy Institute, Union with University Extension College, RUA.

soils, and meteorology all derive from the ideas of the founding fathers, as does work in art (including typography and
graphic communication), music, and languages.

Finances and Buildings

At the beginning the extension college had no endowment,
possessed no freehold property, and had little money at the
bank. Christ Church contributed Mackinder's stipend, and the
fees of students paid some faculty salaries, but the provision
of teaching facilities, a library, and common rooms required
gifts and grants.[18]

In the first session, 1892–93, the college outgrew the Hospitium, and Herbert Sutton bought the adjoining vicarage of
St. Laurence for £4,500.[19] He put up funds for conversion and
agreed that the property could be bought from him, at any time
within three years, for £4,000. The vicarage had a large garden. There, at a cost of £5,000, a building was erected in 1896
to house the Dairy Institute. Again Herbert Sutton advanced
the funds. In the early years Sutton's contributions were crucial, because not everyone locally was eager to offer support.
George Palmer (1818–1897), head of the biscuit firm, did give
£500 in the first year, but he told his third son, Walter, in November, 1892, that he had no confidence the college would
survive.

In January 1896 the college was incorporated and the administration restructured.[20] Under the articles of incorpora-
tion there were established a court of governors, a council, and
an academic board of senior faculty over which the principal presided. The college could now hold property in its own
right. The rooms in the municipal building were given back
to the town council in exchange for a 999-year lease on the
Hospitium at ten shillings per annum. Sutton transferred the
vicarage and the Dairy Institute building to the college and

18. Finance Committee Minute Book, July 1892–December 1895,
    RUA, UCR.
19. Minutes of Council, December 23, 1892, RUA.
20. U.E.C. Memorandum and Articles of Association, 1895, RUA.

agreed to hold a mortgage to allow the properties to be paid for over a period of years.

One major result of the broader administrative structure was to bring in representatives of county councils. Lord Wantage (1832–1901), lord lieutenant of Berkshire, became a member of the court of governors and the council. Wantage was a soldier who had won a Victoria Cross in the Crimean War. By marriage he inherited considerable property. He had a strong sense of duty and a belief that those with wealth should support worthwhile causes. He was a founder of the Red Cross and a landlord who invested in agricultural improvements on a large scale. He was concerned about education in rural areas and supportive of the agricultural work of the college.[21] Reading acquired in Wantage a prestigious patron. He had a strong presence, and when people met him and Lady Wantage they knew they were facing forces for good in the world. Wantage took a long view of events and social needs. He and principal Mackinder understood each other.

Lord Wantage was a key figure in the first major fund-raising drive. The college was still growing rapidly, and it was decided to erect a classroom building in the grounds to link the Dairy Institute and the vicarage. The building cost £9,000, but an appeal was launched for £20,000. Lord Wantage led the way with a gift of £4,000. Following this lead, Walter Palmer subscribed £8,000 and several other Palmer family members gave £500 each. If there had been any doubt that the target would be met, it was dissolved by a master stroke of Lord Wantage. He had been equerry to the Prince of Wales, and he persuaded the heir to the throne to come to Reading to open the buildings on June 11, 1898.[22] Townspeople could not allow their reputation to suffer by failing to meet their target once the prince became involved, and the full amount was raised. Now the college had premises adequate for the next few years, money in the bank, and property assets. The early financial policies of Sutton, Wright, and Mackinder had been shrewd.

21. Harriet Sarah Wantage, *Lord Wantage, V.C., K.C.B.: A Memoir by His Wife.*
22. *Berkshire Chronicle,* June 18, 1898.

When the college wanted to acquire the vicarage, George Palmer had offered to buy it outright, for use by the college, but with the provision that the town council would hold title. At this point Sutton bought the property, freeing the college from the prospect of having the town council as landlord. The decision was sound: around the turn of the century, the council decided it wished to expand the municipal offices onto the college site.

By 1901 it was generally agreed that the college should move out of the town center and south onto the London Road. It was intended to exchange the existing properties for municipal lands, but the town council's lawyers kept finding technical problems. In 1904, after Mackinder had left, Alfred Palmer (1853–1935) broke the deadlock and gave a six-acre site, together with the former Palmer family home, the Acacias, on the south side of London Road, as a new location for the college. It was under the second principal, W. M. Childs, that the college moved to the new buildings away from the city center.

In 1901, after two visits by commissioners, the central government agreed to provide an annual grant of £1,000. In 1902 the name of the institution was changed to University College, Reading. In the space of ten years the tiny extension college had grown to be recognized as one of the thirteen institutions on the University College grant list, along with colleges in such English cities as Bristol, Sheffield, and Leeds.

Students and Curriculum

In the first year the college served 658 students, most of whom were enrolled on a part-time basis. By 1903–1904 there were 1,269 students, and many were full-time. In 1892 the college prepared students for the Department of Science and Art examinations, examinations of the pharmaceutical society, and Oxford Extension certificates. By 1902 the college prepared its students for a diploma describing them as associates of University College, Reading. In addition students were prepared for Cambridge Higher local examinations. After 1899, when the college acquired a status affiliating it with Oxford, students could be entered for responsions[23] at the senior university and, after two years of residence in Oxford, could sit final examinations. Women could be entered directly for all Oxford examinations from Reading. In 1895 Eveline Dowsett passed finals in history by this route, but she was the only one of the successful 109 candidates to reside outside Oxford.

In the first decade many students were sixteen to eighteen years of age. Some were working towards a commercial diploma or certificates for practical skills. Others were prepared for entrance to Oxford, Cambridge, or London University. There were many successes, the most famous being in 1899 when Frank Stenton won a scholarship to Keble College, Oxford. He returned to Reading to become research fellow, professor of history, and vice-chancellor.

In 1899 Reading was recognized as a day training college, and students prepared for teacher certificates issued by the Board of Education. The largest number of certificates was available in agriculture and dairying. In 1894 and 1896 Oxford had conferred upon Reading the power to issue certificates and diplomas in agriculture and related subjects, under the supervision of a joint committee. But a full degree was needed, and in 1898 Lord Wantage and Mackinder tried to persuade Oxford to allow this. Mackinder recalled the meeting dragging on and all of his arguments being "met with palaver." The vice-chancellor of Oxford was sliding away from the issue again when Lord Wantage electrified the meeting by proclaiming, "Vice-Chancellor, I'm disappointed in you!" Oxford vice-chancellors are not often told to smarten up, and Wantage's words had a powerful effect. However, in the end, although the matter went to convocation, Oxford would not relinquish the residence principle: if you wanted an Oxford degree you resided in Oxford for the required number of terms.[24]

The Oxford connection was invaluable in starting the colge, but it did not provide a route to Oxford degrees. Reading ...d reached the point at which it needed the power to grant

23. First of three examinations for the B.A. degree at Oxford University. The examination was abolished in 1960.
24. Mackinder, Notes on the Origin of . . . Reading, March 28, 1926, Mackinder Papers, School of Geography, Oxford.

degrees. In 1899 Lloyd Morgan, principal of University College Bristol suggested that Reading, Exeter, Southampton, and Bristol form a federal university of the West of England. The idea did not develop, probably because Bristol realized that it could get a charter more rapidly on its own.

From 1894, when the Gresham report on the reorganization of London University was published, the idea of taking the college into that university was discussed. It was still being debated in the first decade of the twentieth century, but the distance from central London was too great. The young college had not gone far before it needed a charter to grant its own degrees. The idea of a University of Reading was around even before university college status was attained in 1902. It was very much in the minds of Lord Wantage and Mackinder.

From the start Reading was a regional college that drew students from surrounding counties, particularly to take courses in agriculture. A need quickly emerged for residential accommodation. A register of approved lodgings was compiled and, in 1894–95, two small private halls of residence were established which the college regulated.

In 1901, shortly after the death of her husband, Lady Wantage announced that she wanted to set up a residence for men.[25] The outcome was that she paid for and endowed Wantage Hall, which opened in 1908. The well-proportioned hall was modeled on an Oxford college and possessed a fine quad, a magnificent dining hall, and a residence for a warden. While Lady Wantage's plans were taking shape, Alfred Palmer gave another property, East Thorpe, to form the basis of a new St. Andrew's Hall, which opened in 1911. Reading had a strong residential character earlier than most provincial universities. Again the characteristic was a product of the philosophy of the founding group of Oxford academics and the Wantages.

The Academic Staff

To start a college without a permanent staff and no budgeted positions is a difficult task. However, from the beginning Read-

25. Lady Wantage to Francis H. Wright, July 19, 1901, UAR, UCR 31-2.

ing had excellent faculty members. Oxford was the best academic listening post in the world, and Mackinder was able to hire good people. Some of the early appointments were women, for at that time first-class female scholars had difficulty finding jobs at established institutions and would accept the uncertainties of Reading. For example, Catherine Pollard was born into a Quaker family in Derby in 1869. After successful schooldays at the Mount School, York, she went to Somerville College, Oxford, and studied animal morphology, botany, chemistry, physics, and geology. She got a first at the final examinations in 1891 and then went to live with her parents in Reading. She took a London University B.Sc. degree as well because, until 1920, women could take examinations but not the titles of Oxford degrees. She was engaged by Mackinder in 1892 to teach biology, and physical geography.[26]

Another significant figure was William M. Childs (1869–1939). He was born into an ecclesiastical family in Lincolnshire, which later moved to Portsmouth, where he attended the grammar school. Childs has come to be regarded as the founding father of Reading University, but his early career was ambivalent. As a student at Oxford he was recognized as a historian with a serious purpose, but he did not get a first-class degree when he took finals in 1891. After this disappointment Sadler tested him for extension work, but although he liked Childs's teaching style, he already had more historians than he could employ. Nevertheless Sadler did find Childs a temporary appointment at Aberystwyth University College, and then a position as secretary to A. H. D. Acland, who was a cabinet minister. Childs moved to London and lived at Toynbee Hall, but he left Acland after only a year. Sadler recommended him for the post of part-time lecturer in history at Reading. Childs had been in Mackinder's historical geography class as an undergraduate and, after interview, was offered an appointment.[27] He accepted the job for 1893–94 but contin-

26. Hubert Childs, *W. M. Childs: An Account of His Life and Work*, pp. 87–90.
27. Ibid., p. 35. Report by Reader of Geography for 1889–90, Michelmas Term, 1890, University Archives, SG/R/I/I, Bodleian. Thirty-three students attended the lectures in historical geography. The names

ued to live in Toynbee Hall and obviously looked on Reading as a stopgap. His performance reflected this attitude, and in the summer of 1894 Mackinder told him frankly that he was a disappointment. Childs respected this and produced some good teaching in the following year, but he still vacillated about his future. In the spring of 1895 he was talking of raising sheep in New Zealand.[28]

If Childs could not see his role at the college, those around him did. Mackinder reorganized the literature department and made Childs director, and Michael Sadler donated £100 anonymously to provide a more substantial salary. Henceforth Childs became a key member of the college and began to see in Catherine Pollard qualities that made Reading increasingly attractive. The couple married on July 8, 1897. Catherine continued lecturing until 1899 and then turned to raising four sons. She remained active in developing the hall-of-residence program and in 1899 had been ready to set up a private hall, but the arrival of children forestalled the scheme.

In 1900 Childs was appointed vice-principal; in this capacity he took over the day-to-day management of the college.[29] Mackinder arranged for his own salary to be decreased and that of Childs raised, to take account of changing responsibilities. In 1903, when Mackinder announced he was leaving, there was little debate as to who should be the new principal. Mackinder recommended Childs. This view, widely shared, was quickly acted upon.

If Childs was the organizer, William George de Burgh (1866–1943) grew into the role of academic statesman. De Burgh had a different background from that of many leading figures in the Oxford extension movement. He had been brought up in

include W. M. Childs, the only man from Keble College. The following term there were sixteeen students from Keble, but Childs was not recorded among them. Mackinder's lectures at Oxford and London were to have an influence upon historians, including R. V. Lennard, Frank Stenton, and Lewis Namier. The latter heard Mackinder at LSE and ascribed his appreciation of geographical factors in the shaping of history to Mackinder's teaching (Julia Namier, *Lewis Namier: A Biography*, p. 72).

28. Hubert Childs, *W. M. Childs*, pp. 86–87.
29. Mackinder to Childs, March 7, 1900, Childs papers, UAR.

London, where his father was a barrister and senior civil servant. He was more a product of the establishment than were most of his colleagues at Reading, and he had private wealth. De Burgh was at Merton College, Oxford, from 1885 to 1889. After serving as schoolmaster and extension lecturer, when he came to know Sadler and Mackinder, he joined Toynbee Hall and became friends with Childs. In 1896 he went to Reading, taught Latin, Greek, and philosophy, and became a key figure in the senior common room. He was to be professor of philosophy (1907–34) and dean of the faculty of letters.

There is a tendency for commentators to imply that de Burgh did not get sufficient credit for his work, particularly in the Childs era. This is false. De Burgh occupied exactly the role he wanted. He had opportunities to move, including the chance to start an extension college at Worcester in 1897. He turned this offer aside partly because establishment figures are more comfortable in established places and partly to protect his strong research interests. He published several books, including *Legacy of the Ancient World* (1924), and he had a national reputation as a philosopher.[30]

In 1902, when University College status was attained, the Reading staff was not large by comparison with staffs of similar institutions. However, the quality was relatively high, and many faculty were well known in the academic world, including de Burgh (philosophy), Bowley (economics and statistics), Mackinder (geography), Keeble (botany), Burch (physics), Dickson (meteorology), Frank Morley Fletcher (art), and Gilchrist (agriculture).

Mackinder's relationships with the faculty were good. From the start he fostered a corporate spirit, and by 1897 there was an active senior common room. He did, however, practice an individual form of leadership, in which his vision of the future was paramount. He needed his colleagues to trust him and have faith in his ideas. For a decade he was marvelously successful at "blending dreams and hard sense,"[31] and he pro-

30. A. E. Taylor, "William George de Burgh 1866–1943," *Proceedings of the British Academy* 29 (1943): 371–91.
31. Childs, *W. M. Childs*, p. 11.

vided exactly the type of leadership that the unfunded extension college needed if it were to have a chance of success.

Who were the creators of Reading University? The first thing to say is that the institution is a monument to a phase in which many scholars in Oxford came to believe in the value of extending the work of the university beyond the confines of the colleges. It was from this background that the Oxford extension movement grew.

In 1890–92 Michael Sadler was the key figure who mobilized the interested parties in Christ Church, Oxford Extension, and Reading. Even after Sadler left Oxford in 1895 and went to the intelligence section of the Board of Education, he kept in close touch with Mackinder and visited Reading occasionally. It is difficult to believe that the board's 1899 recognition of Reading as a day training college for teachers did not owe something to Sadler's having suggested opportunities to Mackinder. When Sadler left Oxford Extension, his successor, the historian J. A. R. Marriott (1859–1945), maintained strong links. Marriott frequently offered courses at the college and took an active part in the academic life.

Once the college opened, the major roles were played by Mackinder, Wright, and Sutton in the years 1892 to 1900. Francis Wright lost his wife in an accident shortly after the institution opened, and thereafter he devoted himself to the college. When Reading University received a charter in 1926, Wright became registrar; late in life he was awarded an honorary degree by Oxford and took holy orders. In the early years Mackinder provided the drive, Wright built a solid administrative structure, and Sutton guided the long-term financial strategy. The talents of the men were complementary.

The pace of growth accelerated after 1896, when Lord Wantage took a part in affairs. The years 1900 to 1903 were transitional. Wantage died in 1901, the institution became a University College in 1902, and Childs took over as principal in 1903. Wright continued to carry a major burden, but Sutton's health failed, and Mackinder's role changed. Heavily involved in founding the School of Geography, Oxford (1899), Mackinder was drawn into the affairs of London University, climbed Mount Kenya (1899), contested a parliamentary seat (1900), and published *Britain and the British Seas* (1902).

When Edith Morley came to interview for a language position in 1902, she recorded that Childs, as vice-principal, had taken over the internal management of the college.[32] Mackinder was still active, but he had become minister for external affairs, working on the Boards of Education, Agriculture, and the Treasury for recognition of programs and additional grant monies.

At the beginning of the century Mackinder and Reading had reached turning points. Mackinder, as a result of his work at Oxford, Reading, and London University, was nationally known in educational circles. At forty, poised to take on greater responsibilities, he wanted a role in politics, with the aim of modernizing the English education system. The college had grown from a night school to a University College. A move to a larger site had been decided, another major growth phase was to be embarked upon, and obviously a full-time, resident principal was required. In May 1903, Mackinder told the faculty of his plan to resign, and he indicated some of the thoughts that were influencing him. If he remained he would have to make a commitment to stay for ten years and see the college through to university status, and he wondered if his individualistic style of leadership, which could flourish in the setting of a new college, would be right for the second decade.[33]

Other forces were at work. Many of the original group of enthusiasts in the community had died or been forced to resign from the college council because of ill health or old age. A new body of men wanted to imprint their ideas on the college, and among this group Alfred and George William Palmer were paramount. Mackinder felt that there was an antipathy between himself and members of the Palmer family, that until he went the Palmers would not endow the college. As long as Lord Wantage had been alive, he had made the substantial citizens of Reading feel their civic duty. When Wantage died, the Palmers wished to have the decks cleared, so that they could have an impact on the development of the college. Wantage was a reformer with a gift for seeing far ahead. The Palm-

32. Edith Morley, *Autobiographical Fragments*, RUA.
33. Mackinder to Keltie, March 3, 1891, [1901], RGS (Mackinder misdated the letter); *Berkshire Chronicle*, May 16, May 30, 1903.

ers were no less interested in the college, but they had different qualities. They had been brought up in a tradition of nineteenth-century municipal benefactors. They wanted both to be seen to be giving aid to the town and the college and to be recognized as taking a prominent role in events. Towns expected that of their leading businessmen.

Lord Wantage could say easily that Mackinder was the college's commander-in-chief,[34] but it was not a quality everyone wanted in the principal. Some Reading town councilors thought that the principal's designs were too grand; anyway, shouldn't they be the ones making policy? Mackinder was convinced that he would have been deposed by this element early on except that Christ Church had paid his stipend and replacing him would have cost money.

Within the college Mackinder's leaving was amicable. He and Childs remained friends throughout life. He continued to serve the college in various capacities and used his connections in London to further its interests. But his going was a loss. With hindsight we can see that what was needed was an arrangement whereby Childs took over day-to-day local management and Mackinder occupied a role in which he linked the college with major national opportunities. As it was, the strengths of one man were replaced by those of his successor. Mackinder got far more money and recognition for the tiny extension college, from national agencies, than could have been imagined in 1892. Childs raised amazingly large sums of money in Reading, but he was not linked into the national fabric of educational development in the way Mackinder was.

The first decade of the present century was the decade for giving university colleges charters to grant degrees: Birmingham 1900, Liverpool 1903, Leeds 1904, Sheffield 1905, Belfast 1908, Bristol 1909. In the early years of the century Mackinder enlisted a powerful ally, Lord Haldane, in the cause of gaining Reading a charter. Mackinder and Lord Haldane had known each other, in connection with the London School of Economics, at least since 1902. In October 1906 they went to Reading

34. *Berkshire Chronicle*, May 30, 1903. W. G. de Burgh quoted Lord Wantage.

to open the new buildings on the London Road site. Haldane made the principal speech, expressing the view that it would not be long before the college gained university status. Coming from Haldane this was important, for he was a cabinet minister and a leading advocate of regional universities, and he had been partially responsible for setting up the university grants committee in 1904.[35]

In June 1909 George William Palmer, who had been created a privy councilor in 1906 and must have had knowledge of the chartering of universities, announced that he wanted university status for the college. Together with his brother, Alfred, and Lady Wantage, he put up the funds to carry the project through. But chances were missed, the war intervened, and it was not until 1926 that a charter was granted.

Mackinder's part in the founding of Reading University is among his most important achievements. Reading was chosen for the extension college experiment because he was the Oxford reader in geography. It was smaller than the English provincial university towns, such as Birmingham, Leeds, and Sheffield, which acquired charters before 1914. The next phase of chartering did not arrive until after World War II, when the large cities of Hull, Leicester, Nottingham, and Southampton had their university colleges upgraded to universities. Reading was the only English university to be chartered between the World Wars.[36]

35. Eric Ashby and Mary Anderson, *Portrait of Haldane*, pp. 62–67.
36. J. C. Holt, *The University of Reading: The First Forty Years*. The book deals only with the fifty years from the granting of a charter in 1926. Mackinder gets one mention and a footnote. Holt is critical of Childs and his book, *Making a University*, 1933. This is understandable, since Childs told the Reading story in a very personal way. However, the treatment of Childs and Mackinder is symptomatic of the neglect that the formative period of the University has suffered. The process of forgetting the night-school and adult-education origins was well advanced by the 1930s. Childs retired in 1929, and many other senior colleagues quickly followed him. In December 1935 a new chancellor was installed at Reading and a relatively large number of honorary degrees were given. Childs was recognized, but, although it seems possible that he mentioned Mackinder's name, his influence was waning and Mackinder did not get an award. Mackinder understood what was happening and wrote to Childs on November 30, 1935: "For the good

In terms of size and location, Reading had no reason to expect a university before the 1960s, and even then its proxim-
ity to the colleges of London and Oxford might have mitigated
against such a development. The emergence of a university
at Reading, so early by English standards, was the result of the
vision of Sadler and Mackinder, the determination of Childs
and Wright, and the astonishingly generous level of support
provided by local families, including the Suttons, Wantages,
and Palmers.

---

of the institution the legend of its founding should go down to posterity in right perspective. The story of those first eleven years
ought somehow to be restored to our history. On their scale they
were a very great achievement by you and me—that is not of fundamental importance. But Reading will need the distinction of its
unique history and the College ought not to be written off as mere
prehistory. The University grew not from the day of the Charter
... but from 1892" (Childs papers, RUA).

5

First Ascent of Mount Kenya

Laying the foundations of Reading University took much of Mackinder's time in the last decade of the nineteenth century, but he did not neglect work in geography. In 1893 he helped establish the Geographical Association.[1] The following year he gave the first of an annual series of lectures to teachers at Gresham College, London, with the aim of improving the presentation of geography in schools. He served as president of section E of the British Association in 1895 and became a founding lecturer at the London School of Economics, where he taught commercial geography. Every year he made at least one trip abroad. In 1897 he visited Europe and met leading geographers, including Joseph Partsch, Ferdinand von Richthofen, and Élisée Reclus. Partsch accepted an invitation to

1. In 1892, B. Bentham Dickinson, a master at Rugby School, wrote to the RGS suggesting the establishment of a private association of teachers from fee-paying schools, to make and exchange lantern slides used in teaching geography. The RGS referred the letter to Mackinder, and he convened a meeting at Christ Church on May 20, 1893, where it was decided to form an association for the promotion of geographical teaching generally. Dickinson became secretary of the association, and school masters occupied the other positions on the committee. Dickinson established an elaborate scheme to produce slides of the British empire, and the association also became a pressure group to promote the place of geography in education. Dickinson fell out with several associates involved in schemes to produce slides and maps. The association was reorganized at the annual meeting of January 8, 1900, and opened to all geography teachers, whether in fee-paying schools or not. Dickinson resigned as secretary shortly after this but remained on the committee. By this time the committee was no longer composed solely of school teachers, and Freshfield, Chisholm, and Herbertson were members. Mackinder was not a member of the committee in the early years. He played a part in 1893 and became a vice-president in 1904 (Geographical Association, Sheffield, *Minute Book 1893–1912*).

write *Central Europe* in a series of regional studies that Mackinder had persuaded Heinemann to publish. Mackinder's own *Britain and the British Seas* was to launch the series.

Amidst the teaching, writing, and educational administration he planned an adventure: an expedition to East Africa to climb Mount Kenya and collect materials in the surrounding country. Mackinder was well qualified to undertake the work, being proficient in botany, zoology, geography, and geology. His training with the Rifle Volunteers had provided experience in commanding bodies of men on marches and maneuvers.

The idea of an expedition to East Africa originated in conversations between Mackinder and his Ginsburg relatives, and the Mount Kenya journey became something of a family outing. Although the Mackinder marriage was not a lasting success, partly because of Bonnie's uncertain health, there were enjoyable times. Halford and the Ginsburg family got along well. The Ginsburgs lived in Virginia Water, Surrey, and Mackinder often took the short train ride there from Reading to join his wife for weekends. From 1897 Bonnie lived with her family, and when Mackinder was in Oxford he roomed at the Old Parsonage, at the head of St. Giles, or stayed in Christ Church.

The Ginsburg household was a stimulating place. Dr. Ginsburg had married twice and had five children from the marriages. The four daughters were well educated, and the only son, Benedict Ginsburg, was secretary to the Royal Statistical Society. The family had many friends in the scientific and scholarly world. Hildegarde, Dr. Ginsburg's third daughter was married to Sidney Langford Hinde (1863–1939). Hinde, a Canadian by birth, had trained as a doctor in England and then served in the Congo in the Belgian campaign against the Congo Arabs. In 1894 he undertook exploratory work on the Lualaba River in regions adjoining what is now Zambia and Lake Tanganyika. The exploratory work enhanced Hinde's interests in natural history and ethnology.[2]

2. *Who Was Who, 1929–1940,* "Sidney Langford Hinde" (London: Black, 1941); Sidney Langford Hinde, *The Fall of the Congo Arabs,* and "Three Years' Travel in the Congo Free State," *Geographical Journal* 6 (1895): 426–42.

In 1895, as the British government took over Kenya from the British East Africa Company, with a view to securing Uganda, Hinde joined the civil service and was posted to East Africa. Before Hinde left London he was aware of plans to build the Uganda Railway. The railroad would run from the port of Mombasa, cross the semiarid coastal zone to the highlands of East Africa, pass through the mountainous area that contained Mount Kenya, and terminate on the shores of Lake Victoria. Uganda would be reached by steamers that had been built in Britain, disassembled, transported along the railroad, and reassembled on the lakeshore. The investment, which was not a commercial enterprise, was paid for by the Foreign Office in order to secure Britain's position in East Africa and on the headwaters of the Nile.

Hinde went to East Africa as an administrator, and his province was centered upon Nairobi. Hinde and his wife, Hilda, were to make excellent use of their time in East Africa. They collaborated on a book, *The Last of the Masai* (1901), which Bonnie Mackinder saw through the press. Hilda, following the Ginsburg tradition, worked at languages, producing studies of Masai, Kamba, and Kikuyu, which were published by Cambridge University Press.[3]

Another Ginsburg relative, Campbell B. Hausberg, joined the Kenya project at the planning stage. Hausberg was a fine photographer, a good shot with a rifle, and had enough money to help Mackinder with the costs of the expedition. In addition, Hausberg agreed to be camp master and organized many of the logistical details of the expedition.

Work on the Uganda Railway started in 1896, and Mackinder laid plans for his trip to East Africa. He received some financial help from the Royal Geographical Society, and the Foreign Office gave permission for the expedition to visit Kenya. Mackinder had to improve his field skills. He knew something of surveying and had experience of mountains, but he needed to become more proficient if he was to map in unknown terri-

3. S. L. Hinde and Hildegarde Hinde, *The Last of the Masai* (there is an editor's note, initialed "E. C. M.," by Bonnie Mackinder); H. B. Hinde, *The Masai Language* (editor's note signed "E. C. M."), and *Vocabularies of the Kamba and Kikuyu Languages of East Africa*.

tory and take part in the climb to the summit of Mount Kenya. In the long vacations of 1897 and 1898 he worked at surveying and spent some weeks around Zermatt with a good alpine guide.[4]

The Kenya expedition had all the marks of a Mackinder coup. It involved the bright idea, thought out with a few friends, operationalized on a limited budget, and conceived to take advantage of circumstances which were about to change with the building of the Uganda Railway. Was the whole expedition just a jaunt, undertaken because Mackinder and friends had thought of it? Not at all. Given the limited time and resources available, the expedition was carefully planned and designed to yield topographical and meteorological data in addition to plant and animal specimens. On the other hand, there were professional and personal considerations that gave the expedition more than scientific aims.

The professional objectives were associated with the attitude of some influential members of the Royal Geographical Society. The president, Clements Markham, took the view that real geographers explored unknown lands and brought back new scientific information. Mackinder felt that, if he were to establish credibility with the Markham group, he would have to undertake some exploration. He startled an audience at the Society in 1945, when he received his Patron's medal, by saying as much.[5] However, Mackinder was not simply rising to a bait; Markham, who opposed spending society monies to support geography teaching at universities, was going to withdraw funds from Cambridge, Manchester, and Oxford. Circumstances might be changed if geography was to be taught by an academic who had demonstrated, at least in Markham's view, that he had exploratory experience.

The personal reasons for wanting to climb the mountain were complex. Mackinder was, in a phrase of today, an overachiever. Many of the people he knew were soldiers or explorers who had demonstrated skill and courage in difficult cir-

4. M.P. Auto.
5. *Geographical Journal* 105 (1945): 230–32; Report of Mackinder's speech on receiving the Patron's Medal at the RGS, June 25, 1945.

cumstances. To have worked with Lord Wantage, V.C., the soldier; Markham, the explorer; Henry Bates, the naturalist of Amazonia; Douglas Freshfield, the alpinist; and Mosely, the scientist on *Challenger*, made it difficult for Mackinder not to show that he was capable of similar feats.

Mount Kenya (17,058 ft.) stands in the highlands of East Africa, on the south side of the equator, some three hundred miles inland from Mombasa. The peak, which played an important part in Masai cosmology, was first seen by a European in 1849, when the missionary Johann Ludwig Krapf sighted the mountain. Kilimanjaro (19,340 ft.), the only mountain in Africa to exceed Kenya in height, was climbed in 1889. Kenya was a tougher proposition. Count Teleki had attempted to climb it in 1887, as had the geologist J. W. Gregory in 1893, but both had failed. Gregory (1864–1932), a remarkable character who worked for the British Museum, had been taken to East Africa to give a hunting expedition some scientific credibility.[6] The expedition failed at the coast, but rather than return home without results, Gregory hired porters and set off to explore Mount Kenya and the lakes and rift valleys of East Africa. He made a determined assault on the mountain, cut his way through the forest that extends up to 10,500 feet, and climbed through the high grasslands, reaching the glaciers that surround the peak of Mount Kenya. At around 16,000 feet he was forced to give up when his porter refused to wear boots on the ice. Gregory's expedition revealed the problems of climbing Mount Kenya. Not only was the mountain covered by a snowfield; there also were glaciers to be crossed. The ridges that led upward were not easily connected to the peak. In short, Kenya would not be climbed by a party of keen scramblers. Even good mountaineers would require considerable support and patience if they were to be successful. In addition to his courageous attempt on the mountain, Gregory did fundamental work on the geology of East Africa and advanced ideas on the formation of the rift valleys.

Mount Kenya consists of an extinct volcano. The peaks of

6. *Dictionary of National Biography, 1931–1940* (London: Oxford University Press, 1949.

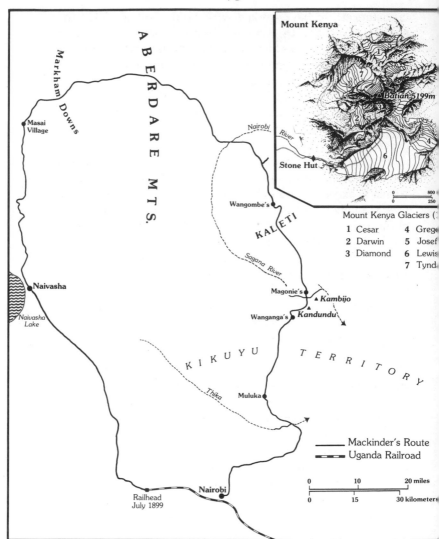

Mount Kenya, Showing Mackinder's Route

the mountain have been weathered out of the igneous rocks
that plugged the vent at depth. The main mass is topped by
three peaks: Batian (17,058 ft.), Nelion (17,022 ft.), and Lenana

(16,355 ft.). The names of the peaks were chosen by Mackinder, following a suggestion by Sidney Hinde that he use the names of Masai chiefs.

In May 1899, when the Uganda Railway reached Nairobi, Sidney and Hilda Hinde moved up to live in a tent at the newly founded town. Mackinder's party left London on June 8, 1899. In addition to Mackinder and Hausburg, Edward Saunders came as a collector and Claude Camburn as taxidermist. Both had been recommended by the Natural History Museum. At Marseilles the group was joined by a guide, Cesar Ollier, and a porter, Joseph Brocherel, both of whom Douglas Freshfield had found for Mackinder on the Italian side of Mont Blanc. Cesar and Joseph were to make it possible to climb Mount Kenya, but even these experienced and determined men did not find the task routine.

The party sailed from Marseilles and arrived at Zanzibar on June 28. In Zanzibar some fifty Swahilis were recruited, including a headman, an interpreter, and eight armed *askaris* (European trained soldiers). The party then moved to Mombasa, arriving on July 4. Now the group began to encounter problems. The coastal area of Kenya was suffering from disease and famine. To avoid the difficulties, most of the party was sent inland by rail to Nairobi. Only Mackinder remained at the coast, making arrangements and persuading a local official that the expedition should be allowed to proceed in time of famine. When Mackinder did reach Nairobi, on July 14, he recorded his impressions of the place: "Nairobi had the aspect of a great common, such as you might find in Surrey. It was bounded north and west by the Kikuyu scrub. Scattered about were a few groups of tents, all that remained of an encampment of three thousand men lately here for a month, who had now gone on with the advance railhead." The only permanent building going up was a bungalow for the chief engineer on the railroad. With help from the district officer, F. G. Hall, Mackinder was able to recruit more porters, and make preparations for the climb.[7]

7. The major documentary sources relating to the Kenya expedition are Mackinder's original diaries, held at Rhodes House, Oxford. The

On July 27, there being smallpox in the area of Nairobi, a camp was set up some miles out of the town. Even this precaution was not sufficient for Hinde. He urged the expedition to move out because, if smallpox broke out in the party, it would not be allowed to travel. Hinde sensed that with problems of famine and disease in Kenya, the Foreign Office would not be enthusiastic about an expedition moving across country. The difficulties around Nairobi put an end to any hopes Hinde may have had of joining the expedition he and Mackinder had planned several years before in England.

On July 28 the march to the mountain started. The party was now nearly two hundred strong, including the Europeans, the men from the coast, the Wakikuyu porters recruited locally, and the Masai guides. For four days they marched approximately northward, across the almost treeless plains. Great herds of game were still evident, and occasionally rhinos charged the column.

On August 1 the party arrived at Muluka, in Wakikuyu country. Mackinder recorded that the southern Kikuyu appeared to be divided into districts, each containing a group of villages. There was no centralized government, and the reception they received varied locally. The land was fertile, and he was impressed by the size of the bananas, maize, beans, and peas, which were the staple crops. The party continued northward and pitched camp at Magonies, close to the igneous hills of Kambijo and Kandundu (4,400 ft.). After accumulating food for three days, they resumed the march, in a northwesterly direction, through land heavily cultivated by skilled agriculturalists who used manure to improve soil fertility. Mackinder wrote: "Here, in the heart of Africa, in a region previously

same repository has a two-volume typed and edited version of the diaries. The School of Geography, Oxford, holds the typescript of a book-length study of the Kenya expedition, based on the diaries. The School also holds photographs, survey, and meteorological data from the expedition. Published accounts include: H. J. Mackinder, "A Journey to the Summit of Mount Kenya, British East Africa," *Geographical Journal* 15 (1900): 453–76, and "Mount Kenya in 1899," *Geographical Journal* 76 (1930): 529–34 (this article contains extracts from Mackinder's Kenya diary). The quote in the text is from Mount Kenya, School of Geography version, July 14, 1899.

approached by half a dozen white men at most, we traversed square miles of standing maize, neatly divided by slight furrows into rectangular half acre plots . . . and we had to pitch camp in a market-place strewn with corn-cobs, or to march for several miles to the next vacant space."[8]

On August 13 the party entered the little country of Kaleti, to the southwest of Mount Kenya, which was ruled by chief Wangombe. Wangombe terrorized the surrounding area, and some of the guides deserted rather than face him. But Wangombe controlled the approaches to the mountain, and it was important to establish good relations with him.

It took two days to march through Kaleti and reach the margins of the forest that clothed the mountain. At a height of 7,200 feet the base camp was established. A body of men was retained to act as food collectors and porters. Others, who were to climb higher on the mountain, were issued with coats discarded by the London police. Now that the party was no longer constantly on the move, fewer men were needed and some porters were paid off.

On August 18, Mackinder and the upward party left the base camp and entered the forest, which, on the southwest side, covered the mountain up to a height of 10,500 feet. Mackinder recorded his impressions of the wooded area: "It was no dark and dank forest . . . such as that of the Congo. The trees were tall but stood apart, letting in light to a rank green undergrowth with abundant flowers, and all the wood was resonant with the song of small birds, but flying insects were strangely rare. I only saw two or three butterflies."[9] As the forest became thicker, the column had to cut a way through. Cesar went first, blazing the trail, and Joseph and the Kikuyu headman worked with machetes to clear creepers. Occasionally an elephant trail made the going easier. The party had some luck getting through the forest. Teleki and Gregory had both taken three days, which had been hard on supplies and on the morale of the porters, who spent a lot of time moving slowly and getting cold. Mackinder's party came through the forest in a

8. Mackinder, *Geographical Journal* 15 (1900): 467.
9. Mount Kenya, School of Geography version, August 18, 1899.

day and camped at its upper edge. On August 20 a substantial camp was set up at 10,300 feet, and the following day they explored the alpine flora zone. Even at this height the footprints of leopards, elephants, and humans were seen, and later, some hunters were encountered by the expedition at 12,000 feet.

On August 23, Mackinder, Cesar, and Joseph reconnoitered the Teleki Valley which was generally thought to be the most promising route to the peaks. But this same day news came up from the base camp that two of the Swahilis had been murdered in an ambush by Kikuyu. Food was not coming into the base camp as arranged, and without additional supplies the expedition would have to be terminated. Hausberg had sent off a party of men to buy food at the government station on Lake Naivasha. The situation was serious, and Mackinder came down the mountain to talk with Hausburg. Obviously food was not going to be obtained regularly from Wangombe's Kaleti, and it was decided to send an additional party, under the command of Saunders, to get more food from Lake Naivasha.[10]

Naivasha was nearly a hundred miles away, on the other side of the Aberdare Mountains. Nairobi was little farther, but Wangombe stood between the expedition and that town. These side journeys set the expedition back, but Cesar and Joseph were left in the Teleki Valley, at 13,000 feet, a mile from the foot of the Lewis Glacier, with instructions to build a stone hut and set up a camp from which the mountain could be climbed. On August 29, Mackinder rejoined them, and the next morning the three men set out for the peaks. They crossed the Lewis Glacier, climbed onto the arete that seemed to lead to the Nelion peak and worked their way upwards. They spent the night at 16,700 feet tied to the rock. The temperature dropped to 29° F, but they had hopes of reaching the peaks, which lay a few hundred feet above them. The next morning, as the party worked up the arete, they came to a deep notch that they could not master. Gregory's view of the mountain was correct: the ridges did not connect easily with the peaks. The first attempt had failed, and the party returned to the

10. Mackinder, *Geographical Journal* 15 (1900): 467.

stone hut beneath the Lewis Glacier. On September 2, Mackinder recorded his feelings: "The high levels are beginning to tell. We are all in low spirits and homesick. Reaction from the rare air, the failure to attain the summit, anxiety about the food caravans and frequent cold feet at night are an obvious explanation. But our nerves are on edge and I hope there will be no quarrelling. The Swahilis are getting a bit difficult. Poor devils."[11] The day after he made this diary entry, Mackinder left the high camp, came to a lower level, got some sleep, and worked on the supply problems. Hausburg took Mackinder's place with Cesar and Joseph. The three made a complete circuit of the peak. Hausburg photographed the mountain from all points of the compass. The professional mountaineers did more reconnaissance and took the opportunity to attack the mountain from the Darwin Glacier. Success eluded them even though they took some risks.

On September 5, Mackinder made the decision that, unless the supply parties sent to Lake Naivasha had returned by September 7, the expedition would have to leave the mountain. Fortunately, on the last day, the Swahili headman, Saunders, and a Captain Gorges, from the government station, arrived with adequate food supplies.

On September 11, Mackinder rejoined Cesar and Joseph at the camp below the Lewis Glacier. The next morning the party of three set out to tackle the mountain again, using a route that combined the climbable sections of the first two attempts. The early part of the route followed the path of the first attempt: an ascent beside the Lewis Glacier before crossing it to the ridge and the next day cutting their way slowly across the Diamond Glacier to the base of Batian. From there it was an easy scramble to the summit, which they reached by noon. They intended to scale Nelion as well, but climbing difficulties prevented this. After the climb the party came all the way down to the stone hut. The last part of the descent was completed in moonlight, and not until 10:20 P.M. did they reach camp. Their feat was not to be repeated for thirty years.

The next three days were spent resting and making ready

11. Mount Kenya, School of Geography version, September 2, 1899.

for another circuit of the peak so that the area could be surveyed using theodolite and plane table. This task was completed by September 19, and the party began the return journey by a route through the Aberdare Range. By this time the porters were becoming demoralized, and some attempted to throw away their loads. One man was punished for jettisoning the frog and bat specimens, and another for discarding botanical specimens.

On September 28 the party reached Captain Gorges's small station on Lake Naivasha. There, awaiting Mackinder was an official letter recalling the expedition! It is difficult to believe that this would not have happened sooner but for the advice and help of Hinde, Hall, and Gorges.

On September 30, Mackinder left, ahead of the group, trying to reach the newly opened School of Geography, Oxford, before the term was half gone. At Nairobi he recorded: "I found everything changed. A galvanized iron town was rising; the hill was already crowned with bungalows; a club had been organized."[12] By November he was back in Oxford and lecturing on a double schedule. The students at Reading welcomed him in triumph, and arrangements were made for a report to the Royal Geographical Society in January 1900.

So the adventure ended. The mountaineering work was a triumph of persistence and determination. For this Cesar and Joseph were rightly given most of the credit by Mackinder. The scientific achievements were substantial given the short time the expedition was in the field. The zoological work added several new species to the known list, and Dr. Bowdler Sharpe of the British Museum reported to the RGS that Mackinder helped to confirm much of the thinking about the general distribution of fauna in highland East and Central Africa. Sharpe named a new species of eagle-owl, brought back from Mount Kenya, after Mackinder. Other species of birds were named after Camburn, Saunders, and Hausburg.[13]

12. Ibid., September 30, 1899.
13. Mackinder, *Geographical Journal* 15 (1900): 478–79 discussion. Reports of the zoological work of the expedition are available in R. Bowdler Sharpe, H. J. Mackinder, Ernest Saunders, and C. Cam-

Because the porters had discarded many specimens, the botanical results were not so rich, but Kew Gardens received several new specimens, including a gladiolus that was named after Mackinder. Everyone had topographical features named for them, and no doubt the "discovery" of Markham Downs was particularly appreciated. More might have been made of the results, but the book Mackinder drafted on Mount Kenya was not published. His mind was moving onto other matters: imperial problems, his future at Reading, and of course the need to make the School of Geography, Oxford, a success. In 1900 he made a foray into national politics and stood as a *Limp* in the election at Warwick and Leamington. Bonnie Mackinder was busy too. She helped organize the Kenya materials and saw the *Masai* book through the press for Sid and Hilda Hinde, who were still in East Africa. Bonnie helped with the political campaign at Warwick and Leamington. However, she was prone to illness, and the extra activity may have helped bring on sickness late in 1900. Mackinder told Keltie of his wife's "dangerous illness" but gave no details.[14] If the illness was stress related, being married to a highly active man—who, in addition to teaching in Oxford, London, and Reading, had led an expedition to East Africa and was embarking on a political career—would not have helped. Late in 1900 the couple made

burn, "On the Birds Collected during the Mackinder Expedition to Mount Kenya," *Proceedings of the Zoological Society* (1900): 596–609; and Oldfield Thomas, "List of Mammals Obtained by Mr. H. J. Mackinder during His Recent Expedition to Mount Kenya, British East Africa," *Proceedings of the Zoological Society* (1900): 173–80. Thomas commented that the mammal fauna of Mount Kenya were practically unknown until the 1899 expedition, Gregory having collected few specimens during his expedition. S. L. Hinde contributed "Remarks on the Mammals Observed during Five Years' Residence in British East Africa" to the same *Proceedings*, pp. 267–68. R. Bowdler Sharpe wrote up *Bubo mackinderi* for the *Bulletin of the British Ornithologists Club* 10 (December 1899): 28. The British Museum acquired forty-five mammal specimens and eighty-five birds from Mackinder (*History of the Collections in the Natural History Department British Museum*, vol. 11 [London: British Museum, 1906]).

14. Mackinder to Keltie, March 3, 1891, RGS. Mackinder misdated the letter, writing 1891 for 1901.

the decision to live apart. There was no divorce. Although it was not obvious at the time, the failure of the marriage was to be the start of a difficult span of years for Mackinder. He spent the Christmas of 1900 in the Senior Common Room at Christ Church.

6

School of Geography

By the late 1880s the Royal Geographical Society had placed teachers of geography at Oxford and Cambridge. However, establishing a new discipline was more complex than hiring two lecturers on short-term appointments. A place had to be found for geography in school curricula, in degree programs, and in the examination systems that gave entry to the civil and military services. Mackinder's appointment as reader in geography at Oxford, in 1887, was the beginning, not the end, of a battle.

In the years immediately after his appointment Mackinder delivered two lectures each week: one in physical geography and the other on the historical geography of a selected region. In the first year attendance was disappointing, but the lectures in historical geography did create interest. Mackinder built on this and formed an alliance with the history faculty. Lectures were delivered at a time convenient for history students, and a place for geographical questions was found in the history examinations. Undergraduates reading for history degrees started to attend in large numbers. Mackinder worked with the natural-science faculty to make similar arrangements, but although he moved his lectures in physical geography to the University Museum, few students attended, there being no examination in the subject.[1]

Mackinder understood that the continuation of his appointment as reader would depend upon the number of students attracted, and the financial support of the Royal Geographi-

---

1. Mackinder to Bates, October 28, 1887, RGS; Mackinder to History Faculty, May 25, 1889, and Report for Academic Year 1890–91, University Archives, SG/R/1/1, Bodleian.

cal Society. In practice the University took the Society's money but, owing to the territorial instincts of faculty members in other disciplines, it was slow to recognize degrees in geography. The subject was not singled out for this treatment. Most of the social sciences—anthropology, economics, and sociology —encountered similar difficulties. Because of the uncertainties, and his desire to work in a broad educational context, Mackinder never devoted all his time to the teaching of geography at Oxford. Between 1887 and 1892 he delivered more extension lectures than lectures to students at Oxford. From October 1892 the duties at Reading occupied most of his time and in 1895 he began to teach at the London School of Economics.

What did Mackinder's geographical teaching consist of at Oxford? Insisting that physical geography was the basis of the subject, he offered courses on the physical geography of the continents and the physical geography of the British Isles. His regional courses were historical. In addition to treatments of the historical geography of the British Isles, the Mediterranean lands, and Western and Central Europe, he lectured on the history of discoveries, stressing the importance of man's changing geographical ideas, not just the chronology of discovery.[2]

For extension lectures the offerings were different. Physiography attracted good audiences. When he was lecturing to schoolteachers, Mackinder usually chose a topic such as "The Relations of Geography to History in Europe and Asia" and "The History and Geography of International Politics." This line of thought culminated in the paper "The Geographical Pivot of History," delivered at the RGS in 1904.

In the early days of his extension work he taught economics and offered courses on "Wealth and Wages," "Revolutions in Commerce," and "The Great Commercial Cities of History." After the opening of the London School of Economics, he lectured on commercial geography and railroads. The material in these courses formed the basis of a series of lectures to the Institute of Bankers in 1899 on the "Great Trade Routes." Although not as well known as the pivot paper, the trade-routes

2. Reports for 1888–89, 1889–90, University Archives, SG/R/1/1, Bodleian.

material is remarkable in the understanding it displays of imperial economic forces. In the opinion of one authority, Mackinder anticipated some of the ideas that J. A. Hobson expressed in his work on imperialism.[3]

In addition to lecturing for university extension, Reading College, and Oxford University, he addressed schools, colleges, and societies at the national and regional level. All these activities took time, and as a result Mackinder did not publish much between 1887 and 1900. The greatest part of his written work was to appear after he reached forty. In retrospect we can see that the founding of Reading University consumed many of Mackinder's potentially productive years as an academic geographer.

At the time the Reading venture was started, Mackinder was working to create a school of geography and trying to retain the interest of the RGS council in the teaching of geography at the university level. This was made more difficult when, in 1893, the Society was split by the "ladies scandal." Women had been admitted to the functions of the Society, but in the early nineties it was proposed to make them eligible for election as fellows. The council agreed to this course of action but then, under pressure from a "dissentist" element led by Admiral McClintock, performed an about-face on the issue. These antics, and the silly things that were said during the controversy, brought about the resignation of some fellows, including Douglas Freshfield. As a result the political balance within the Society shifted. The group that had supported education and research was weakened, in relation to the fellows who wanted the Society to promote exploration and help explorers in the field.

The London School of Geography

Clements Markham, who became president of the Royal Geographical Society in 1893, attempted to undo what he termed the "educational rioting" of Galton, Freshfield, and Keltie. Markham emphasized the need to concentrate and control the

3. Bernard Semmel, *Imperialism and Social Reform*, p. 160.

educational activities of the Society. By this he meant cutting external spending and putting more emphasis on the training of explorers by instructors in the direct employ of the Society. He succeeded in terminating the grant to Owen's College, Manchester, stopping the Oxford grant in 1897–98, and eliminating support for Oxford extension and training colleges.

The first serious shots in the battle were fired in early 1895. The Society wrote to Mackinder stating that: "The original object in respect of the Readership can hardly be said to have been fulfilled. It was hoped that geography would be accorded a substantial place in the university examinations and that thereby the great public schools would be induced to take it up in earnest."[4] Rather than being thrown onto the defensive, Mackinder saw a way to harness Markham's ideas to his scheme for an institute of geography. He took up the notion of concentrating the educational effort of the Society and attached to it a plan to set up a school of geography in London.

Mackinder wrote to Markham on March 6, 1895, telling him that more, not less, would have to be done if geography were to break through the vested interests involved and establish itself in the educational system. He commented, "We want in England something corresponding to the Geographical Institute of Vienna, or at least the less developed geographical department of Harvard University."[5] Mackinder had been elected president of Section E of the British Association, and in his address at the 1895 meeting, in Ipswich, he took the opportunity to review the whole situation. He described the research and teaching work in geography at German universities, suggested that England needed an institute of higher learning in geography, and concluded: "Clearly, if the policy of centralization be agreed to there is only one site for the central school. It must be in London, under the immediate inspiration of that Royal Geographical Society, whose past services to the cause would be a guarantee of support during the early efforts." This paper is sometimes cited as showing Mackinder's

4. Draft Letter to Mackinder, 1895, Mackinder Correspondence, RGS.
5. Mackinder to Markham, March 6, 1895, University Archives, SG/R/1/1, Bodleian.

interest in German geography, but the review of work in German universities was used to demonstrate that England was lagging behind. Like a number of Mackinder's papers this one had elements of a manifesto, and to make sure his point was heard, he had it published three times! It appeared in the *Proceedings of the British Association,* in the *Scottish Geographical Magazine,* and in the *Geographical Journal* as "Modern Geography, German and English."[6]

Having modified Markham's ideas, Mackinder proceeded to voice them in the RGS committees. He started talks between the Society and the newly founded London School of Economics on the possibility of a joint enterprise.[7] Then on March 9, 1896, with the approval of the education committee, he submitted to the finance committee of the Society, a plan for a London School of Geography.[8] A special committee was set up to examine the scheme, and on December 9, Mackinder addressed this group, suggesting that the London School of Geography should be established quickly, before the projected teaching University of London came into being. He argued: "Experience has shown that in older existing universities the vested interests concerned in the other subjects invariably opposed the introduction, or at any rate the full recognition of new subjects. Geographers should, therefore, present a *fait ac-*

6. H. J. Mackinder, Presidential Address, Section E, *Report of the Sixty-Fifth Meeting of the British Association for the Advancement of Science,* p. 474; H. J. Mackinder, "Address to the Geography Society of the British Association," *Scottish Geographical Magazine* 11 (1895): 497–511, and "Modern Geography, German and English," *Geographical Journal* 6 (1895): 367–79.
7. L. M. Cantor, "The Royal Geographical Society and the Projected London Institute of Geography, 1892–1899," *Geographical Journal* 128 (1962): 30–35; Keltie to Secretary, LSE, December 11, 1895, Hewins to Keltie, December 12, 1895, Hewins to RGS, January 6, 1896, RGS; Mackinder to Hewins May 27, 1895, Hewins Papers, 43/212, University of Sheffield. Mackinder accepts Hewins's invitation to lecture on commercial geography at LSE. In accepting, Mackinder specifically mentions his own plans for an institute of geography. Hewins Papers, University of Sheffield, 14/68-101 contains a typescript of Mackinder's plan for a London School of Geography. Clearly Hewins was kept informed of Mackinder's ideas on a school of geography.
8. Committee Minute Book January 1891–June 1897, Education Committee on Proposed School of Geography, June 23, 1896, RGS.

compli to the commissioners who would organize the new university." When the University of London Bill was passed by Parliament in 1898, it contained provisions under which schools could be recognized as teaching components of the university, as had been recommended by the Gresham Commission (1894). When reorganization was carried out in 1900, London University became a teaching university as well as an examining body.[9]

In 1896 the RGS committees had been generally favorable to Mackinder's ideas, but early in 1897 they began to worry about the finances. Eventually, on March 11, 1897, after several long discussions, the finance committee agreed to recommend £400 a year toward the costs of the new London school, provided Mackinder could raise £800 per annum elsewhere.

Efforts were now made to interest the Technical Board of London County Council in the project, and on March 30, 1898, Markham wrote to the board setting out the scheme and asking for financial help. The board, however, wanted to consider how geography was going to fit into commercial education before making a decision.[10] Whether Markham was unhappy at seeing his favorite discipline demoted to the ranks of commercial subjects is not known, but he was a realist. The proposed London School of Geography was disappearing into the bureaucracy of the county council. The Society would have little control over any geography program that might emerge. The London School of Geography died in the spring of 1898.

The London School project had the hallmarks of a classic Mackinder coup. As ideas for organizing a London teaching university emerged, he had seen the early possibilities of creating a new school. The scheme had been maturing for years

---

9. Ibid., Special Committee on the Proposed School of Geography, December 9, 1896.
10. Ibid., Education Committee, March 30, 1898. Markham told the Technical Education Board that the "general intention is to do for geography what the London School of Economics seeks to do for economics." It should be noted that by 1898 the advocates of London University were becoming alarmed at the number of "schools" that were seeking to be recognized within the new university. In Oxford there was talk of establishing a London School of Philosophy and a London School of Botany.

in his mind, but at the chosen time it was presented swiftly and it appeared new and exciting. Mackinder expended immense energy selling the idea and working out the basic organization. The project failed because it lacked financial resources and a second base. Reading was launched by Oxford University Extension, Christ Church, and the local School of Science and Art. The London School was to be launched by the Royal Geographical Society and . . . ?

London University Extension expressed interest but lacked resources, and the London School of Economics was too new and too poor. The Society would pay approximately half the costs but could not find the entire annual budget and did not want sole responsibility. Perhaps if the Society had been established in the present Kensington Gore house, matters might have been different.

Was Mackinder's scheme too ambitious? Probably not. He had been a part of the founding of LSE and was watching his Oxford Extension colleague, Hewins, create a school of economics in the new London University with a few hundred pounds a year. With a little more luck, we might speak of the LSG as easily as we now talk of the LSE.

The School of Geography, Oxford

Before the disappointments over the London School of Geography had been fully realized, all was altered by a change of mind by Sir Clements Markham. Without seeking approval from the council, but using the existing understandings concerning the financing of the proposed London School, Markham wrote privately, on June 16, 1898, to the vice-chancellor of Oxford.[11] He set out a plan for a school of geography and suggested that, in view of accommodation costs in London, Oxford should be the place for the venture. The Society would offer £400 a year if the university would give an equal amount, provide accommodation, and establish a postgraduate diploma

11. Markham to Vice-Chancellor, Oxford, June 16, 1898, RGS. This letter is reproduced in full in D. I. Scargill, "The RGS and the Foundations of Geography at Oxford," *Geographical Journal* 142 (1976): 447.

in geography. The work of the new school would be overseen by a geographical committee consisting of members drawn from the university's Delegacy of the Common Fund and the Society. Rather than seeing his offer of more money to Oxford as a complete about-face, Markham believed he was establishing a new policy. In truth the educational work of the Society was being concentrated and, by the establishment of a geographical committee, the Society did gain a measure of control. Just how Sir Clements had been brought round to advocating increased financial support for geography at Oxford, in the same year he had succeeded in stopping the previous Society grant to the university, is not known. Certainly Mackinder's enthusiasm for the idea of a British school of geography had infected Markham, and as they worked together, Markham's respect for Mackinder apparently had grown, particularly as Mackinder was proving he was the type of geographer who could lead an expedition to East Africa.

In any event Markham's conversion was swift, and it took at least one influential member of the RGS council by surprise. George Brodrick, warden of Merton and a friend of Markham, heard of the scheme in Oxford when he was asked for an opinion on it; he had to write to Keltie, at the RGS, for additional details. When Brodrick learned fully of Markham's plans he wrote, on October 29, offering some solid advice, which went unheeded. Brodrick pointed out that the future of the subject at Oxford would depend "on the place assigned to geography in *examinations.*" He was not hopeful about attracting postgraduate students to take a diploma and felt that an honors degree was required. Markham, and apparently Mackinder, thought that geography would not succeed at the university level until it won a place in the schools, and that would not happen until there were trained teachers in the schools. As there were few school posts specifically for the teaching of geography, perhaps instruction in the subject could be improved by giving some historians, and scientists, additional training and a diploma in geography.[12]

12. Keltie to Brodrick, October 19, 1898, RGS. Scargill, "RGS and Foundations of Geography," reproduces a portion of the Brodrick letter, pp. 447–48.

School of Geography, Oxford,
originally built as the home of the Rev. Mr. Mee in 1899.
Courtesy of the School of Geography, Oxford.

Early in 1899 the university accepted the School of Geography scheme and agreed that a diploma should be created. A geographical committee was established consisting of four members nominated by the Delegacy of the Common Fund, three nominated by the Society, and the vice-chancellor, ex-officio. Mackinder was appointed director and given the tasks of designing the curriculum, engaging a staff, and advertising the new diploma. All this had to be done before June 8, the date of his departure for Kenya.[13]

Staffing the School

The prime new appointment to be made was that of assistant to the reader and lecturer in physical geography of the land. The salary of £270 per annum was just £30 a year less than

13. Minutes of the Committee of the School of Geography, University of Oxford, T.T. 1899 to M.T. 1909, University Archives, SG/M/1/1, Bodleian. The committee had its first meeting on June 8, 1899. Herbertson, Dickson, and Grundy were appointed at the June 8

Mackinder's as director of the School, and the post carried considerable responsibility. The position was offered to Andrew John Herbertson (1866–1915), who was lecturing at Heriot Watt College and helping Bartholomews of Edinburgh with the compilation of atlases.[14] At the time, R. E. Dodge, editor of the *Journal of School Geography,* was suggesting that Herbertson join him at Columbia University in New York. Mackinder knew of this and moved to bring Herbertson to Oxford. The two men had known each other at least since 1895. In the summer of that year they were at the International Geographical Congress, in London, where Herbertson had delivered a paper entitled "The Importance of Geography in Secondary Education." Mackinder had commented lengthily on the topic. In September they were at the Ipswich meeting of the British Association, where Mackinder spoke of the need to establish a school of geography.

Herbertson's training was broad. After working as a surveyor, he went to Edinburgh University and undertook study in physics, mathematics, and geology under Professors Peter Guthrie Tait, George Chrystal, and James Geikie, respectively. Although he did not take a degree, Herbertson had a good scientific background, and in 1891–92 he acted as a teaching assistant to Patrick Geddes, with whom he became closely associated. Henceforth Herbertson's geographical training evolved along two lines. With study at the University of Freiburg, where he was awarded a Ph.D. in 1898, he worked at biogeography and climatology. By contact with Geddes he came to know the work of Frédéric Le Play and Edmond Demolins and began to develop ideas in human geography. The two streams of his training were not distinct, for in this period geography was influenced by plant ecology. Herbertson began to view climatological regions as environments that influence human activity. This aspect of his work can be seen in *Man and His Work: An Introduction to Human Geography* (1899), which Herbert-

meeting, after which Mackinder left for Kenya. Scargill, "RGS and Foundations of Geography," provides more detail on the negotiations between the university and the RGS.

14. E. W. Gilbert, *British Pioneers in Geography,* pp. 180–210.

son wrote with his wife.[15] He had married Frances Richardson, B.A., a teacher at Cheltenham Ladies College, in 1893. Frances shared interests in sociology with her husband and collaborated with him on books.

The appointment of Herbertson to Oxford was important, for potentially it linked Mackinder's historical approach to the more sociological and anthropological view of Geddes. At Oxford, Herbertson and J. L. Myres were to be the representatives of this view, while, at Cambridge, Alfred Haddon (1855–1940) provided an anthropological input to the geography program.

In terms of their interests, Herbertson and Mackinder reinforced each other. Mackinder had been trained initially in the biological sciences, Herbertson in the physical sciences. Subsequently Mackinder had developed an interest in history, while Herbertson had worked at sociology. Mackinder was striving to produce a global picture; Herbertson, in the tradition of Le Play and Geddes, saw local area studies as the starting point for a view of the world, which was rooted in an appreciation of relationships in the home territory. It was the desire of Herbertson to encourage students to understand the social and physical environments around them, that had led him to see geography as a vital subject. He gave much time to the Geographical Association and the campaign to improve the place of geography in secondary education. Here again he augmented Mackinder, whose major links were with the RGS, which had an educational policy aimed at encouraging the growth of geography in higher education.

As administrators the two men complemented each other, at least in their faults. Mackinder was prone to see a little too far ahead; Herbertson was liable to get snared in detail. No doubt observers had some entertaining moments when Mackinder's visionary streak got well ahead of Herbertson's desire to nail down facts. But overall they worked well together.

The part-time position in ancient geography was filled by G. B. Grundy, who had a long connection with Oxford geography, having won the RGS traveling scholarship when it first

15. Frances Herbertson to Geddes, July 30, 1899, Geddes Papers 10536, Scottish National Library (SNL).

was offered, in 1892. The part-time lectureship in the physical geography of the air and ocean was given to Henry Newton Dickson (1866–1922), who lived in Oxford and made a living from teaching, writing, and examining. Dickson, like Herbertson, had studied at Edinburgh University and had been influenced by Chrystal and Tait. Unlike Herbertson and Mackinder, he had not developed an interest in the social sciences. After leaving Edinburgh University, he had gone to work on *Challenger* oceanographic material with John Murray and J. Y. Buchanan. He then took a position at the Plymouth Marine Biological Station and commenced an important study of the circulation of the surface water of the North Atlantic. Dickson had interests in meteorology and was working on the manner in which the temperature and salinity of the sea influenced the character of the air above. He had published an excellent meteorology textbook and could teach surveying.[16] Dickson was a valuable asset, but he had the reputation of being a difficult man to work with. Mackinder knew of this from experience in Reading, having employed Dickson there. Nevertheless Dickson was invited to join the School, and the geographical committee agreed that he should look after administrative matters while Mackinder was in East Africa. This was a mistake.[17]

Mackinder spent more time in Kenya than had been intended. He was late back to Oxford in the autumn of 1899 and could not start teaching until November 7. He rectified the problem by lecturing more frequently, but Dickson had used Mackinder's absence to push himself forward.[18] Having arrived at the School before Herbertson, and being a more forceful character, Dickson worked to ease the assistant to the reader aside, and with some success. Herbertson took time to settle into Oxford, where he found that colleagues "pride themselves

16. *Who Was Who, 1916–1926*, "Henry Newton Dickson" (London: Black, 1929).
17. Minutes of the Committee of the School of Geography, University of Oxford, T.T. 1899 to M.T. 1909, University Archives, SG/M/1/1, Bodleian.
18. Ibid., October 24, 1899. Dickson attempted to get a syllabus accepted in Mackinder's absence, but Mackinder overrode this move at the meeting of November 23, 1899.

on looking backward." But, he observed to Patrick Geddes, his students were "a great joy" and there was life in the place "as well as wine hardened tissue."[19]

In the first two years of the School, course offerings took the following form: Mackinder lectured on the historical geography of North America, Australia, and the Cape, the natural regions of the Old World, the historical geography of Western and Central Europe, and the development of geographical ideas. Herbertson taught courses on the geomorphology of Europe, mountain types, the morphology of the continents, river basins and shore lines, and the geographical cycle. Dickson offered courses on the climate of the British Isles, the physical geography of the sea, climatic regions of the globe, and atmospheric circulation. G. B. Grundy lectured on the historical topography of Greece and the geographical development of the Roman empire. Herbertson and Dickson handled cartography and surveying respectively.

In the summer of 1901 the first diploma examination was held, and four candidates were successful. Now that the School had been running for two years, Mackinder did some stock taking. He was pleased with Dickson's research. His work on the oceanic circulation of the North Atlantic (which led to the award of an Oxford D.Sc.) was important. As a result of a strong recommendation from Mackinder and in spite of a lack of students in physical geography, Dickson's salary was increased. Herbertson's title was changed from that of assistant to the reader to curator of the School and lecturer in regional geography. He then began to offer courses on the regional geography of Europe, America, the British Isles, Africa, and Asia.[20] The intention was to give more regional courses and reduce the amount of staff time devoted to physical geography, where student numbers were limited. The change represented a philosophical shift. In the early months of the School, when Dickson was making an administrative input, a syllabus was adopted that was strong on surveying, measure-

19. A. J. Herbertson to Geddes, December 7, 1904, Geddes Papers 10536, SNL.
20. June 4, 1901, University Archives, SG/M/1/1, Bodleian.

ment, and physics of the atmosphere. The new syllabus represented more of the historical, social, and regional viewpoints of Mackinder and Herbertson.

As part of the 1901 changes, B. V. Darbishire, who had done some work with Friedrich Ratzel in the 1890s, was brought in to help with cartography, and Raymond Beazley (1868–1955) took a part-time position to lecture on the history of discoveries. Mackinder's course load, and his salary, were reduced at his own request.

After an uncertain start, the School was making progress and had six staff members. Most of them were part-time appointments, but the range of expertise was impressive. The academic year 1901–1902 saw more advances. The 1902 Education Act created a place for geography in secondary education, and it was decided that the School should offer summer-vacation courses, at which school teachers would receive some training in geography and teaching methods. The course went well. Thirty teachers took part, all the staff contributed, and Mackinder was particularly pleased with the work of Herbertson and Dickson. Mackinder gave the introductory lecture and one student recorded his impressions: "The wealth of ideas, distinctness of speech, the clearness and precision of the language and, perhaps, above all, the sympathy and infectious enthusiasm of his manner held us interested from beginning to end."[21] Such opinions were not isolated. Mackinder, recognized as one of the best lecturers of his day at Oxford, attracted large audiences for his courses.

The School had developed a distinctive program and some cohesion, but the enrollment in physical geography continued to be poor. This was not a reflection on Dickson; it simply confirmed what was known from Mackinder's experience: natural science students were not interested in physical geography because it was absent from their examination system.

To set against the worry about enrollments was the impressive publication record that members of the School were establishing. In 1902 Mackinder's *Britain and the British Seas*

21. C. C. Carter and C. McGregor, Long Vacation Course, Oxford School of Geography, *Geographical Teacher* 2 (1902): 175.

was published. Some five years previously he had persuaded Heinemann that a new series of books dealing with major regions was needed.[22] He commissioned ten volumes and told his authors that the books were for reading rather than reference and wrote the first volume, on Britain, to show what he meant. Not all the authors delivered manuscripts, but several excellent books were produced, including Thomas Holdich on India, David Hogarth on the Near East, and Partsch on Central Europe.

Britain and the British Seas was an impressive book. It was far better than any other text available, and for British geography it expanded ideas on how to tackle the description of a region. There was no dull catalog of facts but an examination of a series of major themes in a historical context. Hugh Robert Mill reviewed the book in the *Geographical Journal* and called it "new, fresh, and forcible, abounding in unexpected theses handled with skill." Dickson had made Mackinder aware of much recent work in oceanography, and the treatment of the British seas was extensive. The chapters on weather and climate were vividly written. For example, Mackinder described the winter weather conditions when the high pressure of Asia extended over southeast England and produced

> calm, cold, relatively dry weather. A fall of snow may mantle the ground, and a long frost may be inaugurated which yields only with difficulty to the impact of the cyclonic eddies from the west. At such times the heaths and downs of Surrey are bathed in sunshine and crisp, clear air, while mists settle into the Thames Valley, making a silent and bitter gloom continuous through day and night.

The chapters on the physical history of Britain and the rivers gave the first comprehensive account of the geomorphology of the country. The sections on economic geography were excellent, and in the chapter on imperial Britain, Mackinder looked at longer-term problems. He predicted that the European phase of history was passing away and a new balance of

22. Printed circular written by Mackinder March 23, 1897, Mackinder Correspondence 1887–1910, RGS.

power emerging in which a few major states would control affairs.[23]

Around the time Mackinder's book was published, Grundy's *Great Persian War* (1902) appeared, and volume 2 of Beazley's *Dawn of Modern Geography* (1901) was reviewed in the same issue of the *Geographical Journal* as *Britain and the British Seas*. Herbertson was not producing work of the same stature, but he wrote several textbooks at this time and was editing the newly established *Geographical Teacher*. Overall the School was building a reputation by publication within and beyond Oxford.

During the academic year 1903–1904, enrollments at the School followed the established pattern. Mackinder attracted more students than all the other lecturers put together. Beazley and Grundy drew acceptable audiences. Herbertson, as a result of offering a large number of courses, doing much tutoring, and undertaking the duties of curator, made his bread and butter, but Dickson had few students in physical geography or surveying. Finances were tight, and student fees formed a part of the income of the School; the situation could not continue indefinitely.

Mackinder and the geographical committee developed a scheme which would reduce the number of lectures offered by part-timers. In addition, the stipends of Grundy, Beazley, and Dickson were to be partly linked to student numbers and effectively reduced.[24] Grundy and Beazley, who were fellows of colleges, were prepared to accept. Dickson, who had few students and no fellowship at a college, was unhappy. His reaction to the situation was to damage himself, Mackinder, Herbertson, and the School of Geography.

Mackinder's reappointment to the Readership came up early in 1904, and Dickson used the fact that Mackinder had been elected director of the LSE, in December 1903, to get questions raised about his duties and commitments. When the delegates of the Common University Fund considered Mac-

23. H. R. Hill, *Geographical Journal* 19 (1902): 489; H. J. Mackinder, *Britain and the British Seas*, pp. 159–60.
24. May 12, 1904, University Archives, SG/M/1/1, Bodleian.

kinder's reappointment, an inquiry was made as to whether holding the LSE directorship was compatible with the duties of the Reader. The issue was then referred to the geographical committee.[25]

On the surface the question seemed reasonable, but it contained a mischievous element. Mackinder had been principal of Reading College when he started the School, and he had been lecturing at LSE since 1895. These facts were well known, and adjustments had been made to Mackinder's salary to take account of changes in the amount of time he devoted to Oxford geography. The arrangement was not completely satisfactory, but, given that geography was tenuously funded, partially by sources beyond the university, it had been accepted. However, once an issue of this type was raised, other parties entered the battle to advance their view. Within the university were people who doubted the wisdom of creating new diplomas and were uneasy with quasi-departments funded by external bodies, to say nothing of the general reserve at Oxford concerning the validity of such "new" subjects as geography, sociology, economics, and anthropology. In the face of these pressures the geographical committee, putting on a united front, announced that Mackinder had done an excellent job and should remain in charge of the School of Geography.[26] This view was accepted by the delegates of the Common Fund. Mackinder was reappointed to the readership in geography.

Meanwhile Dickson had decided not to teach physical geography in 1904–1905 on the terms offered. He did, however, retain a connection with the School by continuing to teach surveying in conjunction with some military courses. Dickson thus was able to prolong his campaign against Mackinder. Substantial philosophical differences separated Dickson, Herbertson, and Mackinder. Dickson placed a stronger emphasis on the physical basis of the subject than did either Herbertson or Mackinder. In the first years of the present century, Geddes, Herbertson, and Fleure attacked Mackinder for

25. Mackinder to Keltie, Feb. 7, 1904, Myers to Markham, March 7, 1904, RGS.
26. April 28, 1904, University Archives, SG/M/1/1, Bodleian.

what they thought to be his imperial viewpoint. In a letter to Geddes dated December 7, 1904, Herbertson stated, "I am doing my little bit to impress on pupils the need for local patriotism and action . . . to point out how even their imperial patriotism is a middle XIX century conception."[27] In fact Mackinder was not nearly so imperial as Geddes and Fleure suggested, and after the First World War Geddes and Mackinder wrote of the need to preserve provincial identities. There also seemed to be a fear that Mackinder did not like a social-geography approach. It is difficult to know the basis of this concern, for at the London School of Economics Mackinder worked hard to promote sociology and psychology.

The philosophical differences, however, were not the real cause of the disputes. Dickson, ambitious but insecure, tried to usurp the roles of Herbertson and Mackinder. Herbertson, attempting to protect his own position, was not always entirely loyal to the director. The atmosphere at the School was tense, and in the spring of 1905, Mackinder resigned in favor of Herbertson. Even then Mackinder felt strongly that Dickson had to be removed; if past form was any guide, he would work to advantage himself against the new director. Mackinder expressed this view to Keltie at the RGS, and senior council members decided something should be done. When the question of the teaching of surveying came up before the geographical committee, in the summer of 1905, Major Leonard Darwin, R.E., president of the RGS, questioned Herbertson. From this it transpired that, although the new director was prepared to try working with Dickson, he was worried about Dickson's want of ability to get along with colleagues. Major Darwin was a rather unmilitary figure with a reputation for fair dealing. To protect Herbertson he insisted that Dickson should not be employed to teach in the coming year.[28]

27. A. J. Herbertson to P. Geddes, December 7, 1904, Geddes Papers 10536, SNL.
28. Mackinder to Keltie, January 6, 1905, RGS; Gwen Raverat, *Period Piece: A Cambridge Childhood*, pp. 195–96; Report by L. Darwin on Meeting of Oxford Geographical Committee, May 30, 1905, RGS.

Dickson was disappointed by all of this. He had ambitions to succeed Mackinder in the readership. Apparently he did not understand that in academic palace revolutions the "hatchet man" rarely succeeds to office. Dickson went on to become professor of geography at Reading in 1907 and president of the Royal Meteorological Society, and he did excellent work on the Admiralty handbooks in World War I. His going was an academic loss. If he had continued to work at Oxford, he would have been able to bring meteorology into a central place in British geography much earlier than was the case. Dickson was an able scientist with drive, ambition, and administrative skills. Had he been patient, he might have succeeded Mackinder and become the second director of the School of Geography. He should have known that there was not long to wait, for it was obvious to everyone at Reading, in particular, that eventually Mackinder would move to London and involve himself in politics. Mackinder wanted a few more years in Oxford to firmly establish the School. Dickson destroyed that plan and greatly weakened the School as a result.

The School was not the same after 1905, and it nearly died with Herbertson in 1915. One important factor concerning the standing of the School in the university was that it did not have an Oxford man at the helm. Many in the university were critical of geography and other new subjects, but they respected Mackinder for his achievements. He had a first-class honors degree in natural sciences and had been elected to a college fellowship. Such successes have always been highly respected in an academic community that prides itself on the quality of the competitions it runs. Such conservatives as Charles Oman might view Mackinder's commitment to extension work and geography as overdone, but they did not doubt his intellectual quality. Herbertson was an outsider with a strange history. He had no undergraduate degree but possessed a doctorate from a German university. At the time such higher degrees were suspect in Oxford. It was Oman who engineered the denial of a chair to Herbertson at the first attempt.

There is nothing particularly abnormal about the disagreements that took place at the School. Such upheavals are common in academia. People are worried about promotion, tenure, recognition, salary, and prospects. Until recently, with the establishment of grievance procedures, very little of academic fights has been recorded. In the case just described, because a university and an outside body were involved, several people had to write letters; as a result, much of the story can be pieced together in the archives of the RGS. After the row, the usual silences were preserved. Mackinder and the RGS adopted the stance that he had resigned in 1905 to take on the directorship of LSE—a position he had held since 1903. Dickson said he had left Oxford because of the unsatisfactory terms offered. Later he excluded any mention of the School of Geography from his entry in *Who's Who*. Grundy, who had been involved, never mentioned the School in his autobiography, *Fifty-Five Years at Oxford* (1945).

The damage done to British geography and the Oxford School was massive and lingering. The School lost two substantial scholars, who were not replaced. In 1906 Herbertson was doing most of the teaching in physical and regional geography, but although he had a great range of knowledge, there is no substitute for exposing students to a variety of approaches. By normal standards Herbertson did good work. He built up student numbers, particularly after 1907, when geography was given a place in the civil-service examination system. He enlarged the summer-school program and attempted to start an institute for the study of colonial territories. After one denial he was awarded the personal title of professor. Herbertson's teaching and administrative load was heavy, and in addition he did considerable work for the Geographical Association and the *Geographical Teacher*. He published many texts but was unable to find the time to develop his ideas in human geography. He died of heart failure in 1915 at a relatively early age. His wife, who had collaborated with him on so many projects, died a few days later, and their son was killed in the First World War.

At Herbertson's death there was no obvious successor other than Dickson, who held the chair of geography at Reading. However, even a new generation of senior men at the RGS was not impressed by aspects of his personality, and his name did not receive serious consideration.[29] Mackinder's opinion was sought. His view was that, although Dickson was prone to intrigue in a subordinate role, he might be a good head of a department. Mackinder was, however, in favor of appointing a reader with specialization in historical rather than physical geography. Eventually H. O. Beckit, assistant to Herbertson since 1908, succeeded to the readership. In 1932, after Beckit's death, the university established an honors degree and appointed the royal engineer, Lt. Col. Kenneth Mason, superintendent of the Survey of India, as professor. The first of the English universities in modern times to establish courses in geography was just about the last to offer an honors degree in the subject. Part of the explanation for the delay lies in the debilitating battles of 1904 and 1905.[30]

Ironically, while the dispute was coming to a head in early 1904, two members of the Oxford School delivered papers at the RGS that were to be among the most influential in British geography during the first half of the twentieth century. In January 1904 Mackinder delivered his famous paper on "The Geographical Pivot of History." The following month Herbertson presented his work on the world's "Major Natural Regions," which was to have a powerful influence on the way geographical knowledge was organized, written about, and taught.[31] The papers were illustrative of the real and potential strength of Oxford geography. Once the school was split, neither Herbertson nor Mackinder produced such a contribution again.

29. Hinks to President, October 22, 1915, RGS.
30. Mackinder to Freshfield, April 7, 1919, RGS. Lt. Col. Kenneth Mason, M.C. (1887–1976), was educated at the Royal Military Academy, Woolwich. In 1906 he was commissioned in the Royal Engineers and joined the Survey of India in 1909, rising to be superintendent. Mason's distinguished work in the Himalayas was recognized by the RGS with the award of the Founder's Gold Medal in 1927. He was professor of geography at Oxford and fellow of Hertford College from 1932 to 1953.
31. A. J. Herbertson, "The Major Natural Regions: An Essay in Systematic Geography," *Geographical Journal* 25 (1905): 300–12.

7

The Geographical Pivot of History

Halford Mackinder is most widely known for the ideas he presented in a paper entitled "The Geographical Pivot of History."[1] The paper was read at the Royal Geographical Society on the evening of January 25, 1904. It carried a novel message. The world was coming to the end of the "Columbian epoch," and the importance of sea power was declining relative to landpower. Railroads were providing the means of integrating continental areas, and Mackinder suggested that there was a pivotal area, "in the closed heart-land of Euro-Asia," which was likely to become the seat of world power. In 1904 the potential of the pivot area was balanced by the states in the inner, or marginal, crescent. However,

> the offsetting of the balance of power in favour of the pivot state, resulting in its expansion over the marginal lands of Euro-Asia, would permit of the use of vast continental resources for fleet-building, and the empire of the world would then be in sight. This might happen if Germany were to ally herself with Russia.[2]

Mackinder did not make specific predictions as to the countries that might try to gain control of the pivot, for he wanted

Portions of this chapter have appeared in B. W. Blouet, "The Maritime Origins of Mackinder's Heartland Thesis," *Great Plains–Rocky Mountain Geographical Journal* 2 (1973): 6–11, and in B. W. Blouet, "Halford Mackinder's Heartland Thesis: Formative Influences," *Great Plains–Rocky Mountain Geographical Journal* 5 (1976): 2–6.
1. H. J. Mackinder, "The Geographical Pivot of History," *Geographical Journal* 23 (1904): 421–44, reprinted in H. J. Mackinder, *The Scope and Methods of Geography and the Geographical Pivot of History.* The reprint has an introduction by E. W. Gilbert. For an excellent discussion of criticisms of Mackinder's strategic ideas, see W. H. Parker, *Mackinder: Geography as an Aid to Statecraft.*
2. Mackinder, "Pivot," p. 436.

to bring out underlying geographical realities. His contention was that, from a geographical point of view, the pivot area would be at the heart of any struggle for world dominance. The implications of Mackinder's statement were immense. If he were right, the basis of British influence in the world—sea power—was about to be diminished. The position of Britain as a world power would become tenuous, and the future of the empire insecure. Read in conjunction with his published lectures to the Institute of Bankers, the paper amounted to a suggestion that Britain faced eclipse as a leading power militarily, industrially, and imperially.

The boldly conceived Pivot paper offered a view contrary to the conventional wisdom of the day, which held that sea power was essential if a country wished to be a world power.[3] The paper contains original elements, though in part it had grown out of an historical experience, and there had been previous, rudimentary, statements of Mackinder's idea. The paper was a product of its time and a national experience, in the same way that Frederick Jackson Turner's "frontier thesis" reflected widespread American feelings in the 1890s.[4]

Since the sixteenth century Britain had enjoyed an expanding importance in the world and had become a leading commercial and military force on the basis of sea power. Until well into the nineteenth century, ships were the means by which most goods could be moved over long distances from areas of production to markets. Britain, and the other mari-

3. Gerald S. Graham, *The Politics of Naval Supremacy*, p. 29. P. M. Kennedy, *The Rise and Fall of British Naval Mastery*, contains an excellent discussion of the views of Mackinder and Mahan. William E. Livezey, *Mahan on Sea Power*, pp. 286–89, discusses Mahan in relation to Mackinder and contends that Mackinder had a strong influence on German geopoliticians. Robert Seager II, *Alfred Thayer Mahan*, pp. 462–63, suggests that Mahan's *The Problems of Asia* (1902) contains a form of the heartland concept. S. B. Cohen, *Geography and Politics in a World Divided*, discusses Mahan and provides an excellent overview of Mackinder's strategic views as expressed in 1904, 1919, and 1943.
4. James C. Malin, "Space and History: Reflections on the Closed-Space Doctrines of Turner and Mackinder and the Challenge of Those Ideas by the Air Age," *Agricultural History* 18 (1944): 65–74, 107–26; reprinted in *History and Ecology*, ed. Robert P. Swiergenga, pp. 68–84.

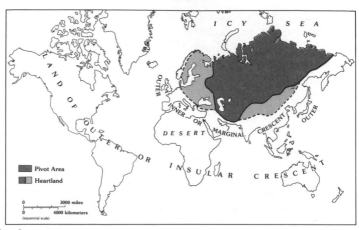

The Geographical Pivot of History and the Heartland

time powers, controlled the sea routes over which goods moved and therefore held most of the wealth and power. The land-oriented states, though potent in the vicinity of their own territory, were relatively unimportant on the worldwide scene.

During the late eighteenth century and early nineteenth century, because of its early lead in the industrial revolution, Britain was able to build a navy that, for a period at least, could not have been challenged by all the other navies of the world, even had they been able to act in concert. This British naval force, exploiting the unity of the ocean, lent such effective support to imperial policy that maritime power appeared to be the sole key to world dominance.

Inevitably the British lead was reduced. Other countries had their industrial revolutions and began to close the gap. What was more, industrialization created modes of transportation that made it practical to move goods and people long distances overland. Railroads made it possible to unify land areas and integrate their resources for use by larger states than previously had been possible. It was at just this time, when the balance was shifting to the continent-sized land power, that Mackinder was writing.

Mackinder's theme had been exercised by other writers in

Britain. In 1882 Professor J. R. Seeley, in lectures at Cambridge University, had told his audience:

> Now if it be true that a larger type of state than any hitherto known is springing up in the world, is not this a serious consideration for those states which rise only to the old level of magnitude? Russia already presses somewhat heavily on central Europe: what will she do when with her vast territory and population she equals Germany in intelligence and organization, when all her railways are made, her people educated, and her government settled on a solid basis — and let us remember that if we allow her half a century to make such progress her population will at the end of that time be not eighty but nearly 160 million. At that time which many present may live to see, Russia and the United States will surpass in power the states now called great.[5]

The question of size disadvantage had been widely understood in Britain since 1870.[6] At the first Imperial Conference (1887) the matter formed part of the arguments for a federated British empire, capable of matching the might of the new powers that were emerging from the unification and integration processes in Germany, Russia, Italy, and the United States. As larger powers emerged, the problems of the British began to take new forms. The British home fleet, which had been capable of holding in check the European navies, was now liable to be outflanked. The problem first developed in the Mediterranean, where France was using her frontage onto that sea more effectively and Russia was exerting pressure on the Straits leading from the Black Sea into the Mediterranean. By the latter part of the century, Germany was a unified power leapfrogging beyond the established European arena with such schemes as a Berlin-to-Baghdad railway that might well have embarrassed the British, not only at the straits, but eventually in the Persian Gulf and the Indian Ocean. Constantly the Russians were perceived to be working toward the enlargement

5. J. R. Seeley, *The Expansion of England*, pp. 300–301.
6. J. A. Froude, "England and Her Colonies," *Fraser's Magazine*, New Series 1 (1870): 16.

of their domain: pressing upon Turkey and Iran and absorbing or gaining influence in the various Central Asian states that bordered the Indian empire.

As the land powers of Europe became progressively better organized in an easterly direction, so the British imperial response migrated eastward. Construction of the big naval dockyards at Malta began in the 1840s, Cyprus was leased as a base from Turkey in 1878, and Alexandria was built up as a naval station after the occupation of Egypt in 1882. The Indian Ocean had long been a center of British interest, and apart from the military machine on the subcontinent, there were bases at Simonstown, Mauritius, Aden, Trincomalee, Penang, and Singapore, together with a naval presence in the Persian Gulf that had been established by the Bombay Marine, the naval arm of the East India Company.[7] With one exception, all the bases had been established to meet maritime threats or to ensure maritime control. Only Aden, taken in 1839, had resulted from a landpower threat: the thrust of Mehemet Ali into Arabia from Egypt, a move that was seen as endangering the imperial route to India.

After the announcement of the proposed Berlin-Baghdad railway, new arrangements had to be made to strengthen the defense of British India. In 1899 the Sheikdom of Kuwait entered into a protectorate relationship with the United Kingdom in which the sheik undertook not to cede, sell, or otherwise alienate any part of his territory save with the consent of Her Majesty's government. This move went a great way towards stalling any possible German and Russian ideas of making naval dispositions at the head of the Persian Gulf. Other sheikdoms on the Arabian shore entered into similar agreements in the same period. The Iranian coast of the gulf was secured from outside influence by the Anglo-Russian agreement of 1907, which defined spheres of influence for Britain and Russia in Persia (Iran).[8]

The elongation of imperial power continued. The building

7. Gerald S. Graham, *Great Britain in the Indian Ocean, 1810–1850.*
8. Briton Cooper Busch, *Britain and the Persian Gulf, 1894–1914,* p. 113.

of the Trans-Siberian Railway in the 1890s and the early years of the twentieth century, together with the growth of Japanese might, necessitated the strengthening of military facilities at Singapore and Hong Kong. The Japanese, who it was felt might provide a maritime balance to Russian land-based Asiatic power, were bound to friendship by the Anglo-Japanese Alliance (1902).[9]

Adroit as the diplomatic and military activity was, it could not indefinitely prolong British advantage. During the period of maneuver, however, many politicians, diplomats, and military men had come into contact with the problems of operating along the peripheries of a great landmass in competition with powers possessing comparatively compact lines of communication and holding the additional advantage of being able to switch the real or feigned pressure points along British imperial routes with comparative ease.

In the late nineteenth century the question of how Britain was to deal with the continental powers, principally Russia, in relation to the security of the empire was a prime problem of foreign policy and one on which there was a great range of opinion. There were, however, two major viewpoints. One group viewed Russian dominance of the Asian landmass as inevitable and incontestable. A representative of this school of thought was Lord George Hamilton, Secretary of State for India (1895–1903), who held that the construction of railroads in Central Asia had enabled Russia to concentrate forces in such strength that it would be difficult for British naval power to act as a counterpoise. In 1900 Hamilton even went so far as to suggest that the Russian "advance is like that of a glacier, slow but omnipotent."[10]

The leader of the opposing school was Lord Curzon, Viceroy of India (1898–1905). Curzon knew his regional geography firsthand and stated his view of matters in 1899 in the following terms: "I will no more admit that an irresistible destiny is going to plant Russia in the Persian Gulf than in Kabul or

9. C. A. Fisher, "The Britain of the East," *Modern Asian Studies* 2 (1968): 343.
10. David Dilks, *Curzon in India*, vol. 1, p. 132.

Constantinople. South of a certain line in Asia her future is much more what we choose to make it than what she can make it herself."[11] It is probable that Curzon and Mackinder had an influence upon each other's thinking. They were about the same age, and their careers at the Oxford Union were in part coincident. Both were members of the Royal Geographical Society, and Curzon was president from 1911 until 1914. Curzon served as Foreign Secretary from 1919 to 1924 and was to send Mackinder to South Russia as the British High Commissioner.

At the British Association meeting at Newcastle in 1889, which Mackinder attended, Curzon read a paper on the "Central Asian Railway in Relation to the Commercial Rivalry of England and Russia."[12] Although the paper was of limited scope, Mackinder was already thinking about the Russian problem, and Curzon's ideas certainly would have registered in that context. It would be surprising if Mackinder was not familiar with Curzon's books on Asian problems, which were published in the early 1890s. The traffic in ideas, however, was not only in one direction. In 1904, when Curzon was Viceroy of India, he read with interest the Pivot paper and commented favorably upon it to Scot Keltie.[13]

The Pivot paper was written at a time when the issue of imperial defense was under debate. The Boer War (1899–1902) opened up the question of Britain's position in the world. A large British army was engaged by a guerrilla force in the Boer territory of South Africa. Some of Britain's European rivals viewed the situation with much satisfaction. The Russians proposed a continental coalition in favor of the Boers, and the Kaiser took public pleasure in exploiting Britain's problems. As a result of the war a reevaluation of defense problems was

11. Earl J. Ronaldshay, *The Life of Lord Curzon,* vol. 2, p. 99.
12. George N. Curzon, "Central Asian Railway in Relation to the Commercial Rivalry of England and Russia," *Report of the Fifty-Ninth Meeting of the British Association for the Advancement of Science,* p. 663; *Russia in Central Asia in 1889 and the Anglo-Russian Question; Persia and the Anglo-Persian Question; Problems of the Far East;* and *Frontiers.*
13. Andrew Goudie, "George Nathaniel Curzon: Superior Geographer," *Geographical Journal* 146 (1980): 207.

undertaken and an Imperial Defence Committee set up in the hope of coordinating defense activities in the Empire.

In 1943 Mackinder discussed in *Foreign Affairs* the origins of the Pivot paper and recognized that the Boer War had played a part in his thinking.[14] But the roots of the idea went deeper. In Mackinder's early childhood the only danger to Britain's position in the world was in Asiatic Russia, where a threat to India was perceived. In September 1870 he became aware of another power on the international scene. Walking past the post office on his way home from Gainsborough Grammar School, he learned, from a telegram pinned to the door, that the Prussian army had defeated the French at Sedan. Other defeats were to follow, including the fall of Paris in January 1871, before a peace was signed in March of the same year. The Franco-Prussian War must have been a topic of interest in the Mackinder household. The family governess, Mme. Hosteller, was French, Draper Mackinder had lived in Paris, and several relatives on the Hewitt side had experience of Germany.

Mackinder may have carried the germ of another idea away from Gainsborough. The little port on the River Trent had enjoyed something of a trade boom until the opening of the Manchester, Sheffield, and Lincolnshire Railway in 1849. In the decades after the coming of the railroad, Gainsborough went into decline as goods were carried directly, by rail, to such seaports as Grimsby. Gainsborough had its competitive position changed by the railroad. The theme of the economic impact of railroads was carefully worked out in Mackinder's lectures to the Institute of Bankers in 1889, and the strategic impact of railroads was a fundamental part of the Pivot paper. As Mackinder put the matter in 1899:

> The first effect of the railway was simply to accentuate what had been taking place, and to emphasize the world's center as essentially oceanic. The explanation of this lies in the fact that the railways were, in the first instance, simply local railways, which fed the ocean traffic. . . . The great oceanic trade retained its essential mechanism; but we are now just beginning to see

14. H. J. Mackinder, "The Round World and the Winning of the Peace," *Foreign Affairs* 21 (1943): 595–605.

what is probably a reversal. . . . It is necessary to emphasize certain characteristics of the railway traffic as compared with the steamship traffic. For instance, it is open to doubt whether, in ordinary consideration of the subject, sufficient attention has been given to the influence of railways on the development of German trading. Germany occupies a central position on the Continent of Europe. It is possible for Germany to send goods, without breaking bulk, from the factory to any market on the continent which is accessible by railway. Owing to the abolition of Customs Frontiers within Germany itself, owing to the system of sealing wagons and sending goods across countries in bond, it has come about that the railway has a very significant advantage over the steamer, in that it is possible to avoid even that breaking of bulk which is involved in taking goods from the factory down to the ship. There can be little question that in Italy, in Hungary, and in the Near East, Germany must have an advantage over England, for the simple reason that we send our goods by rail down to the port, after which they have to be handled and placed on the ship. They are carried round to the port at the far end, and there they have to be handled again and placed on the railway, and only then carried to their destination; whereas, by a system of private sidings, it is possible for a factory in Germany to place the goods actually upon the truck within the factory gates, for them to be sealed by the Customs authorities, and to be sent from there, let us say, to Madrid, to Naples, or to Constantinople, and to be, in some cases, actually run into the yard of the wholesale dealer who is to distribute them within the town. That is a very important fact in connection with the development of the railway system, the full consequences of which we have yet to see.[15]

When the Pivot paper began to crystalize in Mackinder's mind, he was in contact with Sidney and Beatrice Webb and a group that had connections with the London School of Economics. In 1902 the Webbs founded the Co-Efficients dining club. The members included Sir Edward Grey, a future foreign secretary, Lord Haldane, a future minister of war, Leo Maxse,

15. H. J. Mackinder, "The Great Trade Routes," *Journal of the Institute of Bankers* 21 (1900): 153–54.

editor of the *National Review* and a correspondent of Admiral A. T. Mahan, Bertrand Russell, H. G. Wells, and L. S. Amery, who became a prominent member of the Conservative party. There is no doubt that this powerful group stimulated Mackinder's thinking and that several of Mackinder's ideas found their way into political circles via the Co-Efficients. Amery and Mackinder interacted well. Amery never forgot the message of the Pivot paper, and he was to be instrumental in bringing Mackinder into the imperial unity group of Lord Milner. Sir Edward Grey, and particularly Haldane, found Mackinder's views interesting, and both became members of the Liberal government that signed the Anglo-Russian agreement of 1907 —a move that can be seen as an effort to support Russia in the Pivot area against possible German expansion.

Did Mackinder really have the intellectual power to stand out as a man of ideas in the brilliant Co-Efficients group? In Amery's opinion the answer was yes. In *My Political Life* he described Mackinder as having a more forceful personality and a more powerful brain than either Grey or Haldane.[16]

H. G. Wells was an invigorating influence who sparked Mackinder's imagination. In 1902, Wells published a book entitled *Anticipations*, which speculated on long-term scientific and social developments in the twentieth century. Some of the predictions were remarkably accurate, although the book erred in many respects, for example, in underestimating the pace at which air travel would become commonplace, and the potential for Russian development. One of the ideas Wells advanced in *Anticipations* may have played a part in the evolution of the Pivot paper. Wells believed that eventually a world state would emerge: "Through whatever disorders of danger and conflict, whatever centuries of misunderstanding and bloodshed, men may still have to pass, the process nevertheless aims finally, and will attain the establishment of one world-state at peace within itself."[17] It would be a small step for the geographer to ask, "Where is the likely seat of power

16. L. S. Amery, *My Political Life*, vol. 1: *England before the Storm, 1896–1914*, pp. 228–29.
17. H. G. Wells, *Anticipations of the Reactions of Mechanical and Scientific Progress upon Human Life and Thought*, p. 245 (originally

of such a state?" For Mackinder the answer was the interior of Eurasia. Should that area come under the control of one power, or alliance of powers, "the empire of the world would then be in sight."

The Co-Efficients dining club met on a monthly basis to discuss imperial problems and promote the cause of national efficiency. On April 27, 1903, the Co-Efficients discussed the question "What should be the relations of Britain to the great European powers?" From the minutes taken by Amery and Mackinder it seems that the group thought that differences with France were minimal. This is not surprising, given that the Anglo-French Entente was to be signed in 1904. On the other hand it was felt that both Russia and Germany presented threats:

> On the whole, all were agreed in recognizing the Russians' persistent and probably inevitable advance against the weak states of Asia as the problem of international politics which was fundamentally the most difficult. But several urged that it was a problem that did not demand early hostilities, and that in shaping our international friendships and alliances it might be postponed to the most urgent opposition to German ambitions.[18]

In short the group had difficulty in deciding whether Germany or Russia was the problem. A few months after the discussion Mackinder wrote the Pivot paper and conceptually overcame the choice of Germany or Russia as the major threat by suggesting that in the heart of Eurasia there was a strategic area which, if controlled by one power or alliance, would give that force long-term advantages.

At the turn of the century, Britain was greatly alarmed by Germany's determination to build a high-seas fleet for the specific purpose of gaining power in world affairs. The German "Navy Laws" of 1898 and 1900, which were seen as a direct threat to British naval supremacy, led to fears that Germany

---

published as a series of articles in the *Fortnightly Review* during 1901).

18. The Co-Efficients produced printed minutes as a record of their discussions. One set of minutes is in the H. G. Wells papers at the University of Illinois.

would use the new navy to invade Britain. In 1903 Erskine Childers published a widely read novel entitled *The Riddle of the Sands*. Childers made frequent reference to Mahan's theories in his story, which was about two young Englishmen on a sailing vacation uncovering a German plot to invade Britain. Mackinder did not mention the threat of German sea power directly in the Pivot paper, but the message was clear. The real danger was not that Germany would immediately challenge British naval might in the West. However, if Germany were to push eastward, take over the pivot, and use the "vast continental resources for fleet-building," Britain was likely to be overwhelmed in the long run.

In the spring and summer of 1903 British politics was polarized by Joseph Chamberlain's speech of May 15, which called for a policy of tariff reform to replace the British tradition of free trade. Mackinder had fought the Warwick and Leamington constituency in the election of 1900 as a Liberal Imperialist. With moral support from Amery, Mackinder resigned from the Liberal Party and became a "Chamberlainite" in 1903.[19] Against this political background the Pivot paper is not entirely an academic exercise. If the logic of physical geography was leading to the possible emergence of a major power dominating Eurasia, then the best course of action for Britain was to promote a policy of imperial unity, within a preferential tariff structure, in order to have sufficient size and economic weight to remain a great power.[20]

It is sometimes suggested that the Pivot paper attracted little notice beyond the Royal Geographical Society. This view is incorrect. The paper was widely reported in leading newspapers including the *Times* of London and the *Glasgow Herald*. It was discussed in such monthly and weekly journals as the *National Geographic* and the *Spectator*. The *Spectator* did not see how a "great and wealthy pivotal state" could be formed

19. Julian Amery, *Life of Joseph Chamberlain*, vol. 5: *Joseph Chamberlain and the Tariff Reform Campaign*, p. 184; Mackinder to Lyttleton, October 16, 1903, Chandos papers, Chan 5/17, Churchill College, Cambridge.
20. Bernard Semmel discusses this point in *Imperialism and Social Reform*, p. 164.

on the resources of northern Asia. It doubted that rail transport was significantly cheaper than carriage by water, and it suggested that territorial expansion would bring problems because "pivots suffer from attrition." In any event, some new factor would always appear to upset the long-range calculations of the type Mackinder was making.[21] In fact most of the basic criticisms of the Pivot idea were made around the time the paper was presented. In the discussion at the RGS, immediately after the reading of the paper, Amery suggested that, in view of the coming of air power, the type of strategic land power versus sea power analysis presented by Mackinder might soon be outdated. The Wright brothers flew on December 17, 1903, and the Pivot paper was delivered on January 25, 1904; thus, Amery was quick to see the prospect of air power.[22]

The Pivot paper was not ignored, but what impact did it have? From the evidence of Eyre Crowe's famous 1907 Foreign Office memorandum, on the importance of naval power to Britain, we might conclude that Mackinder's prophecy regarding the decline of sea power had little effect. Crowe declared:

> The general character of England's foreign policy is determined by the immutable conditions of her geographical situation on the ocean flank of Europe as an island State with vast oversea colonies and dependencies, whose existence and survival as an independent community are inseparably bound up with the possession of preponderant sea power. The tremendous influence of such preponderance has been described in the classical pages of Captain Mahan. No one now disputes it. Sea power is more potent than land power.[23]

21. *The Times*, January 26, 1904; *Glasgow Herald*, January 26, 1904; "The Geographical Pivot of History," *National Geographic* 15 (1904): 331–35; *The Spectator*, January 30, 1904. The *Times* story ran: "A generation ago steam and the Suez Canal appeared to have increased the mobility of sea power relative to land power. Railways acted directly as feeders to ocean-going commerce. But transcontinental railways were now transmuting the conditions of land power, and nowhere could they have such effect as in the closed heart-land of Euro-Asia."
22. Mackinder, "Pivot," discussion, p. 441.
23. G. P. Gooch and Harold Temperley, eds., *British Documents on the Origins of the War 1898–1914*, vol. 3, p. 402 (London: HMSO).

However, Crowe went on to recognize that British sea power could be overthrown by "a general combination of the world," and he was particularly alarmed about German expansion resulting in a subservient position for Britain. Crowe advocated alliances for Britain that would align her against the strongest European power—Germany. The Crowe memorandum had a strong impact on the foreign secretary of the day, Earl Grey.

The whole system of international understandings, built up by Britain in the first decade of the century with Japan (1902), France (1904), and Russia (1907), was part of an effort to ensure that no one power would dominate the Eurasian landmass. In this sense Mackinder's paper was very much a part of a national debate on defense. There are direct links between Mackinder and such key figures in the debate as Spenser Wilkinson, Haldane, and Earl Grey. Wilkinson heard the paper read and, as he was married to Eyre Crowe's sister, he may well have carried the pivot message to the influential Foreign Office official.

Over the long term the Pivot paper, which encapsulated a part of a national experience and viewpoint, was in the main line of thought that led to the idea of Britain's joining a European common market. In 1934 Arnold Toynbee published *A Study of History,* which contained the opinion that Europe would be overwhelmed by surrounding giant states, an idea that is in the same tradition as the Pivot paper. A future prime minister of Britain, Harold Macmillan, read Toynbee, and the idea became a part of his arguments for attempting to take Britain into the European Common Market in 1961.[24] A more direct route for the idea to enter into Conservative Party thinking was provided by Amery. Whatever his criticisms of the Pivot paper, he never forgot Mackinder's presentation, and, when he became an advocate of a united Europe, he quoted Mackinder in support of his views. In a communication to General Smuts on December 15, 1943, Amery wrote:

> The whole of the west and south European group by themselves are in the long run incapable of defence in the event of Russia

24. Harold Macmillan, *At the End of the Day, 1961–1963,* pp. 1–2.

and Germany coming together—the great danger of civilization which Mackinder predicted years ago. It is I think essential for the future that Germany, after due punishment and relative weakening, should be brought back into the main European fold, and that a European commonwealth should comprise the main European block of countries up to the Russian border. It is precisely from that point of view that I am anxious that continental Europe should constitute a definite world economic group. . . .[25]

By 1943 Winston Churchill was making powerful statements about the need to create a United States of Europe. Churchill and Mackinder were associated over Britain's Russian intervention, in 1919, and it is likely that Churchill was presented with a copy of *Democratic Ideals and Reality.* This is not to suggest that the line of thought concerning the need for a United Europe was Mackinder's invention. Mackinder was part of a long tradition of British thought that understood the danger of Eurasia's coming under the control of one power or alliance. In this tradition, Mackinder made the clearest statement of the problem, which tells us something of the vision of the man and his intellectual capacity.

25. Jean Van Der Poel, ed., *Selections from the Smuts Papers,* vol. 6, pp. 469–72.

London School of Economics

The creation of the London School of Economics and Political Science, in 1895, is sometimes portrayed as a special event, engineered by Sidney Webb.[1] However, the foundation of LSE is part of a broader picture involving the establishment of the social sciences in higher education and the creation of university extension colleges. It is hardly surprising, therefore, to find many of the academics associated with Oxford University Extension appearing among the founding fathers of the London School of Economics.

At the Oxford meeting of the British Association in September 1894 the economists discussed the position of their subject in British universities. Economics, like geography, lacked a full place in the curriculum and had a low status in the civil-service entry examinations. Sidney Webb was at the Oxford meeting, and the discussion had an impact upon him. Within a few weeks he was sure of the Hutchinson bequest to the Fabian Society and decided that it should be used to start a school of economics in London rather than be frittered away on political activities. Webb thought that Graham Wallas, an Oxford graduate who was active in London County Council politics, would be director of the School, but Wallas turned down the opportunity and the position was promptly offered to W. A. S. Hewins. Hewins, another of Michael Sadler's young

1. Standard works on the London School of Economics and the personalities surrounding it include F. A. Hayek, "The London School of Economics, 1895–1945," *Economica*, New Series 13, no. 49 (1946); Sir Sidney Caine, *The History of the Foundation of the London School of Economics and Political Science*; Norman Mackenzie, *The Letters of Sidney and Beatrice Webb*; Norman Mackenzie and Jeanne Mackenzie, *The Fabians*; Lisanne Radice, *Beatrice and Sydney Webb: Fabian Socialists*.

lecturers, taught economics and other subjects for Oxford Extension. By now we can predict the characteristics of an extension lecturer, and Hewins was no exception. His origins were provincial: he was born at Wolverhampton in 1866 and educated at Wolverhampton Grammar School. He had religious convictions and was to embrace the Roman Catholic faith later in life. Hewins made his way to Oxford by scholarships and, after taking a degree, stayed on in an effort to pioneer one of the "new" subjects—economics. But he was rebuffed. At the beckoning of Sidney Webb he went to London in 1895 to start LSE and created what became, in 1898, a faculty of economics within the reorganized University of London.[2]

According to the first LSE prospectus, which Hewins had printed in Oxford, the School would have as its special aim:

> the study and investigation of the concrete facts of industrial life and the actual working of economics and political relations as they exist or have existed in the United Kingdom and in foreign countries. With this object in view the School will provide scientific training in methods of investigation and research, and will afford facilities to British and foreign students to undertake special studies of industrial life and original work in economics and political science.[3]

2. *Report of the Sixty-Fourth Meeting of the British Association for the Advancement of Science* (London: John Murray, 1894), p. 737. Some Webb-Hewins correspondence is reproduced in Norman Mackenzie, *Letters of Sidney and Beatrice Webb*, vol. 2, pp. 28–34; Webb to Hewins, March 24, 1895, 43/127–139, HP: "Graham Wallas now decides after all that he cannot undertake the Directorship! This is an unexpected blow. I write in haste to ask whether it would be possible for you to come and see me any day during the coming week. We should be very pleased to put you up. It is now a matter of serious import whether the scheme can be carried through. I am still keen on it, and if it should be possible for you to help to a greater extent than we contemplated it might still be done." Hewins telegraphed his acceptance of the invitation on March 29. Webb wrote, after a meeting of the Hutchinson Trustees, offering Hewins the directorship (S. Webb to Hewins, March 29, 1895, 43/134, HP), which he took up on April 1, 1895.
3. W. A. S. Hewins, *The London School of Economics and Political Science*. A. Kadish, *The Oxford Economists in the Late Nineteenth Century*, suggests that LSE was founded, academically, by a group of Oxford scholars who dissented from prevailing economic opinions.

At first LSE offered courses only in the evening, and the staff was drawn largely from among the extension lecturers at Oxford and Cambridge. At the beginning, Mackinder was brought in to teach commercial geography. All the lecturers adopted the university extension format: a lecture was delivered, and then there was a break followed by discussion and the return of essays. Initially students were not prepared for a specific examination. It was intended that the courses be useful preparation for civil service, Institute of Bankers, and commercial-education examinations.

Within a few years some four hundred part-time students were registered for evening courses. The London County Council, via the Technical Education Board, was providing substantial financial support, and funds also were provided by the Lincolnshire landowner Hickman Bacon and the London Chamber of Commerce. Student fees were low and provided only a limited part of the income. Generally finances were not a problem, and the deficits that had to be met out of the Hutchinson bequest were small.

The School began in rented property but, in 1902, moved into new premises at Clare Market. The British Library of Political and Economic Science was established in 1896, students started to be prepared for London University degrees in 1898, and LSE became a recognized part of the University in 1900. After this recognition the School received a portion of the London County Council grant to London University, in addition to the existing funding made available by the Technical Education Board. In 1901 the School was incorporated. The LSE grew with extraordinary rapidity in the early years, in conditions which favored growth. London University was being organized as a teaching institution, and it was possible to get the social sciences included without too much opposition from established subjects. Sidney Webb and Hewins took the opportunity brilliantly.

The work of Hewins at LSE deserves a reassessment. He has suffered from some of the widely published diary entries that Beatrice Webb made concerning him.[4] It must be remem-

4. Beatrice Webb, *Our Partnership*, ed. Barbara Drake and Margaret I.

bered that Beatrice thought of the School as a creation of her husband. In some senses it was, but the story is more complex. In the view of Miss C. S. Mactaggart, secretary to the School under the first three directors, Hewins "worked far too hard and had too much to do."[5] LSE could not have been a success without an immense amount of enthusiasm and commitment on the part of the first director and the small staff, most of whom stayed with the School for many years. Sidney Webb produced funds from the Hutchinson bequest, the London County Council, and his influential friends. Hewins built the staff and the academic program.

In May 1903, Joseph Chamberlain called for a program of tariff reform, which is to say protectionist policies. Hewins became heavily involved in the debate, resigned from LSE on November 18, 1903, and at the end of term went to head the Tariff Commission that Chamberlain set up. Within days of the Hewins resignation, Sidney Webb moved to appoint Mackinder to the directorship. In terms of the experience Mackinder had gained, particularly at University College Reading, there probably was no better-qualified candidate in the country, and Mackinder had the added advantage of having worked closely with the Webbs and the School since its inception.[6]

Halford Mackinder was not primarily known as an economist, but he had studied the subject at Oxford under Bon-

Cole, p. 204. In some senses LSE has suffered from having its chronicles recorded as Webb elitist history. Not nearly enough is made of the early contributions of such distinguished scholars as Sir Arthur Bowley, Edwin Cannan, Leonard Hobhouse, and Herbert Foxwell, to say nothing of the work of Hewins and Mackinder.

5. Recollections of Miss Mactaggart, Feb. 11, 1933, *Material on the History of the School*, R (S.R.) 1101, British Library of Political and Economic Science (BLPES).

6. W. A. S. Hewins, *The Apologia of an Imperialist*; Senate Minutes, December 16, 1903, London University; Letter from Hewins (November 18) announcing his resignation; Letter from Sidney Webb (November 29) proposing Mackinder as director of LSE and Mackinder appointed at £300 per annum. The salary of the director of LSE was paid by London University. Mackinder became the appointed teacher of economic geography at London University by action of the senate on May 28, 1902. The appointment carried the responsibility to advise on the organization of the Geography Department of the School of Economics.

amy Price and taught economics and economic history for Oxford Extension and the University of Pennsylvania. As Bernard Semmel has pointed out, Mackinder had a grasp of Britain's economic situation. Familiar with the arguments of the free traders and the tariff reformers, he pointed to a fundamental dichotomy in the economic structure of Britain: the interests of the British banking and commercial world were not the same as those of the manufacturer. The bankers gained advantages from free trade, but the manufacturers stood to benefit from the protective policies advocated by the tariff reformers.[7]

These themes were worked out in a series of lectures that Mackinder delivered to members of the Institute of Bankers, in London and Cardiff, late in 1899. The lectures appear to have been delivered without a prepared text but were taken down, transcribed, and then printed in the *Journal of the Institute of Bankers*. The lectures were not extemporaneous. Many of the themes are to be found in synopses of Mackinder's extension lectures published in the early 1890s. But speaking without notes, Mackinder lectured in a fresh and lively style. It is easy to understand why he was regarded as one of the best lecturers of his day.[8]

Mackinder discerned a series of chapters in British economic history in a global context. The search for seaways to the Indies, by Portugal and Spain, had developed new routes to the Indian Ocean and to the Americas. Among the main beneficiaries of these developments were the Dutch, who distributed Portuguese goods in northern Europe. Amsterdam grew to be the major commercial center of Western Europe. In the seventeenth century, London and the English replaced Amsterdam and the Dutch at the heart of the commercial

7. Bernard Semmel, *Imperialism and Social Reform: English Social Imperial Thought, 1895–1914*, pp. 157–60.
8. H. J. Mackinder, "The Great Trade Routes," *Journal of the Institute of Bankers* 21 (1900). The four lectures in the series were published separately. The epitome of the first lecture appeared in January, pp. 1–6. Lecture 2 appeared in March, pp. 137–46, and lecture 3, pp. 147–55. Lecture 4 was published in May, pp. 266–73. At the conclusion of the course was an examination. The questions were published in April, pp. 220–21, together with a short examiner's report by Mackinder.

system. The British became traders and bankers before they became manufacturers.

In the closing decades of the eighteenth century and in the early nineteenth century the industrial revolution in Britain produced new forms of manufacturing activity. There followed an intensive phase of development, linked to the opening of the temperate grasslands, in which capital and manufactured goods were exported from Britain, while foodstuffs and raw materials were imported to the manufacturing centers. In equipping the new settlement areas, the British manufacturing industries grew greatly in importance, and there was a maximum differentiation of economic activity on a global scale. But this was a passing phase. The building of railroads across the grasslands allowed the dispersal of manufacturing industry into areas that formerly had produced the raw materials and foodstuffs. Mackinder suggested that the effect of this dispersion of industry would be to lessen the relative importance of British manufacturing on the world scene. However, the whole economic system still required a center that supplied capital and acted as a clearing house. The importance of London as a financial center was likely to increase. Thus British banking would gain in importance, but British manufacturing would decline, at least relatively. Mackinder concluded:

> This gives the real key to the struggle between our free trade policy and the protection of other countries—we are essentially the people with capital, and those who have capital always share the proceeds of the activity of brains and muscles of other countries. . . . It is for the maintenance of our position in the world, because we are the great lenders, that we have been driven to increase our Empire.[9]

The arguments developed in the Institute of Bankers lectures were not elaborately worked out, but as Semmel has suggested, Mackinder was anticipating the ideas of J. A. Hobson in *Imperialism* (1902) and of Joseph Schumpeter in *Imperialism and Social Classes* (1951). He had gone to the heart of the different interests that were likely to be involved in the pro-

9. *Journal of the Institute of Bankers* 21 (March 1900): 155.

tection debate that broke out in Britain in 1903.[10] In 1899 Mackinder was still a member of the Liberal party and supported free trade. Within a few years he was advocating protectionism and imperial-preference trade policies. Put another way, when the great political schism opened up in the first decade of the twentieth century, Mackinder abandoned the interests of the City of London financiers and sided with the provincial industrialists, such as the Marshalls of Gainsborough, who made agricultural equipment.

Directing the School of Economics and Political Science

In 1903 the London School of Economics had assets, income, and a recognized path to London University degrees for registered students. But it was still a night-school operation. There were few full-time students, and the academic program needed broadening to include more social sciences. Mackinder was given the chance to build the School in these areas. At Reading the academic program constantly ran ahead of resources and status. At LSE, degree status and resources had been put in place, but the program needed to be expanded.

For Hewins, resigning the directorship of LSE must have been a relief. He was a man of considerable energy, but he undertook much of the detailed management of the School. Mackinder's style was different. He looked at the big picture and allowed his small administrative staff to run day-to-day affairs. The Court of Governors had less detailed business brought before it and was required to consider wider issues and long-term plans. A finance and general-purpose committee, established in 1903, looked after routine business.[11]

10. Richard A. Rempel, *Unionists Divided: Arthur Balfour, Joseph Chamberlain and the Unionist Free Traders*, pp. 98–104, discusses the types of economic interests that tended to back the free-trade and protectionist positions. Dilwyn Porter, "The Unionist Tariff Reformers, 1903–1914," p. 21, suggests that Semmel overstates matters by suggesting that city interests were all against tariff reform.
11. Sources on LSE include: Calendars, London School of Economics and Political Science, 1895– ; Court of Governors Minute Book, 1905–1909; Director's Annual Report, 1904–1908; Finance and General Purpose Committee Minute Books, 1903–1906, 1906–1910; and *Clare Market Review*, 1905–1908.

Mackinder's most important task was to bring the School fully into the life of the University of London. This was essential if the academic program were to grow and the School to receive a full share of the resources that the University disbursed. On accepting the directorship Mackinder was elected by the faculty of Economics (largely the staff of LSE) to represent it on the Senate of London University. As a member of the senate, Mackinder served on a variety of university committees, including: commercial education, the council for external students, Galton's eugenics committee, and the academic council. By his own count, Mackinder served on nearly forty university committees during his term as director, and he spent more time in South Kensington than at Clare Market. The time was well spent. As Miss Mactaggart put the matter "under Mackinder we ceased to be so much at enmity with the other colleges of the university, an intercollegiate system of attendance at lectures was worked out." It was typical of Mackinder to spend his time building external relationships and to leave the people at LSE to get on with their own work. Much the same thing had happened at Reading.[12]

When Mackinder took over at LSE, virtually all courses were offered at night, and the students were largely part time. Day classes were introduced by the simple mechanism of getting faculty to deliver the same lecture twice in the same day. For example, Lilian Knowles, the excellent historian whom Mackinder appointed, offered "Outlines of English History" in midmorning and then again in the early evening.

In terms of curriculum, Mackinder concentrated on enlarging the range of social sciences and encouraging postgraduate work. Shortly after he became director, Mackinder was able to get himself onto the university committee responsible for advising on the use of the Martin White benefaction. White was a wealthy Scot with a wish, and the means, to promote the study of sociology. As a result of Mackinder's influence, the funds were steered towards LSE and became the basis of

12. Senate Minutes, January 27, 1904, London University; C. S. Mactaggart, Recollections of the London School of Economics and Political Science, 1942, R (S.R.) 1101, BLPES.

a strong sociology program. E. A. Westermarck, the Finnish social anthropologist, and A. Haddon of Cambridge offered the early work in sociology and ethnology. One intention was that members of the Indian Civil Service attend courses and be better able to understand the cultural groups they would encounter overseas. In addition the teaching of geography was strengthened, with help from the RGS, and efforts were made to offer work in psychology.[13]

Some of the largest advances were made in the development of programs to serve executives in the railroad, banking, and insurance industries. Mackinder established committees, with representatives of leading companies, to advise upon work in railroad and insurance subjects. The railroad committee proved invaluable. The companies got the courses they wanted and, in the process, came to understand the needs of the School. One result was that the railroad industry contributed generously to LSE finances. Insurance work was less successful. When the insurance institute was established, teaching of the subject was reduced.

Attempts to offer advanced training to business executives and civil servants were a part of the School's mission to promote "national efficiency." Although the phrase was not in use in the 1880s and 1890s, the work of university extension lecturers had been in pursuit of this goal, at least in the sense that improved education standards were felt to have significant economic results. Extension lecturers provided the majority of the staff at LSE in the early years. Around the turn of the century the slogan "national efficiency" was widely used, and several of the leading advocates became associated with LSE. Haldane became a member of the Court of Governors, and in 1901 the Earl of Rosebery became president of the School. Rosebery was a leading figure in the efficiency movement and for a time was thought likely to launch a new political party that supported social reform, national efficiency, and imperial unity.[14]

13. Senate Minutes, July 5, 1905, London University; Finance and General Purpose Committee Minutes, February 17, 1908, Records in Connaught House, London School of Economics (LSE).
14. G. R. Searle, *The Quest for National Efficiency*, pp. 107–41.

The interest in national efficiency did have an influence upon the work of the School. In addition to courses for civil servants and executives, in 1906, Haldane (minister of war) and Mackinder worked out a scheme to give army officers administrative training. The first groups of officers entered the School in January 1907 and completed a six-month course in which they were exposed to accounting, law, economic theory, geography, statistics, and transportational studies. The program was a success and continued for many years. The War Office paid the full cost, which was a great help to the finances. Playfully the student magazine, *The Clare Market Review,* dubbed the officers the "Mackindergarten." The course gave the School a body of full-time students, and it became possible to open up a refectory for the benefit of all staff and students.[15]

In 1907, at the third attempt, the School was admitted to the Treasury List of universities and colleges that received direct government grants.

Teaching and Writing

In addition to directing the School and taking an active part in the government of the University of London, Mackinder carried a heavy teaching load. In the academic year 1906–1907, for example, he offered an introductory geography course (eleven lectures), regional geography (nineteen), historical geography (ten), a map class with A. J. Sargent (nineteen), and a course entitled "British History from a Geographical Viewpoint" (ten). The average attendance at the last-named course, nearly two hundred students, was the largest enrollment at the School. As at Oxford, Mackinder's skills as a lecturer on geographical themes in a historical context were recognized as supreme and attracted a broad range of students.

The years as director of LSE (1903–1908) covered Mackinder's most productive time as a writer and editor. The Regions of the World series continued to demand work, and a revised

15. H. J. Mackinder, *Address delivered on 10 January, 1907, on the occasion of the opening of the class for the Administrative Training of Army Officers,* pp. 1–12; Finance and General Purpose Committee, 1906–1910, November 29, 1906, LSE.

edition of *Britain and the British Seas* was issued in 1907. Around the same time he began to publish a series of elementary texts, starting with *Our Own Islands* (1906) and *Lands Beyond the Channel* (1908). These volumes helped provide adequate materials to support geography teaching in schools. In 1907 he spent the summer with his brother, the Reverend Lionel Mackinder, at Wotten-under-Edge and wrote *The Rhine: Its Valley and History* (1908). Mackinder edited the section on Britain for the *Encyclopedia Americana* and obtained contributions from such associates as the Lyttletons and the Webbs. He wrote on contemporary problems in various journals, including the *National Review*,[16] continued to write geographical articles, and produced materials on Britain and India for the Colonial Office visual-instruction committee.

The books produced in this period enjoyed good sales and brought in a flow of royalties. Mackinder may have hoped that the income would allow him to give up administration and devote his time to political causes. But although the writing of educational books produced useful additional income, he did not derive a living from the activity.

The years as director of LSE, after he had given up the Oxford School of Geography, were a partially settled period in Mackinder's life. The failure of the marriage was fading, and he was no longer commuting between Oxford, Reading, and London. He was comfortably housed in St. James Court and had access to Speldhurst Rectory, in the heart of the Kentish Weald, where he retreated to walk, think, and write. The administration of LSE did not consume him in the way Reading had done. At no time during his tenure at Reading did the college have a secure financial base. Mackinder, and then Childs, carried the constant burden of having to find money to run the institution. At LSE things were different, and Mackinder found the School easy work after Reading. Sidney Webb thought Mackinder ran the place with two fingers of one hand, or so he commented to William Beveridge when persuading him to take on the directorship.[17] But no sooner were there signs of set-

---

16. H. J. Mackinder, "Man-Power as a Measure of National and Imperial Strength," *National Review* 45 (1905): 136–43.
17. Lord Beveridge, *The London School of Economics and Its Problems,*

tling down than his love of beginnings drew Mackinder away to start new careers in business and politics.

Social and Political Life

At LSE Mackinder had the opportunity to meet a variety of politicians and people in national affairs. The Webbs maintained friendships and connections with a range of public figures. A major purpose of the School, from their perspective, was to provide a center that could contribute to social and economic debates and influence the decision-making process. From the foundation of LSE there were gatherings at which the Webbs introduced the staff of the School to their friends and associates. Mackinder took part in the social life, and by the spring of 1902 he had made enough impact on Beatrice Webb for her to find him worthy of one of the short character sketches that make her diary interesting reading:

> Mackinder is an able lecturer on commercial geography, energetic traveller and organizer: he has political ambitions and is by way of attaching himself to the Rosebery group. He is a coarse grained individual (Bertrand says *brutal*) but with a certain capacity for oratory and strong picturesque statement. If he got his foot on the ladder he might go far towards the top: especially as there is an absence of able young men. Signs in him of negroid blood?[18]

In the summer of 1902 Mackinder joined the Webbs for a vacation in the Cotswolds, and Beatrice commented, "We have had pleasant companionship with Hewins, Mackinder and Russells." It was at this time that the Co-Efficients dining club was born. Mackinder recalled in later life:

> One day in that holiday, as I rode beside Mrs. Webb, she told me there were certain men of her acquaintance whom she would like to bring together, and that she thought of founding a dining club, from which she would retire after presiding at the in-

---

1919–37, p. 65. Beveridge tells the same story in *Power and Influence*, p. 168.
18. Diary of Beatrice Webb, transcript, June 1902, BLPES.

augural dinner. Her husband, who was riding just ahead of us, threw back over his shoulder—"I will give your club its name —the Co-Efficients."

The neat title reflected the interest of the Webbs in national efficiency and collective solutions to social and economic problems. The projected dinner was held on November 6, 1902, and the club formally organized with a dozen members consisting of: L. S. Amery; Carlyon Bellairs, R.N.; Clinton Dawkins (banker); Sir Edward Grey, M.P.; R. B. Haldane, M.P.; Hewins; Mackinder; L. J. Maxse; W. P. Reeves (New Zealand High Commissioner); Bertrand Russell; Sidney Webb; and H. G. Wells. Amery and Mackinder were elected as secretaries. All members had interests in the Empire and social reform, with the exception of Russell, who resigned in June 1903 and was replaced by Michael Sadler.[19]

After the first dinner in the Webbs' home, and the second at Haldane's, the group met about once a month at St. Ermine's Hotel, Westminster, or at The Ship, in Whitehall. The general format was that after a good dinner one member introduced a topic, general discussion followed, and the secretaries made a précis, which later was printed and distributed to members.

At the first working dinner, on December 8, 1902, Pember Reeves introduced a discussion of closer political relations within the Empire. In Janury 1903, Hewins spoke on preferential Empire trade. There followed, at subsequent meetings, Bellairs on imperial defense and Dawkins on relations with America. The discussion on America resulted in Amery's advocating, in the event of Britain's going to war against the United States, mobilizing the bulk of the population and shipping it across the Atlantic to defend Canada! The argument precipitated the resignation of Bertrand Russell. In later meet-

19. Diary of Beatrice Webb, July 21, 1902, BLPES; M.P. Auto. A similar account of the group's naming is to be found in Mackinder to Sidney Webb, May 3, 1943, Passfield 11 4 n. 150, BLPES. For accounts of the Co-Efficients, see Bernard Semmel, *Imperialism and Social Reform*, pp. 62–73, and Robert Scally, *The Origins of the Lloyd George Coalition: The Politics of Social Imperialism, 1900–1918*, pp. 73–95.

ings of the 1902–1903 session, Grey spoke on relations with European powers and Amery on imperial developments.

The following session, 1903–1904, was more concerned with social reform in Britain: Webb on minimum-living standards, Wells on municipal enterprises, Dawkins on national service, Mackinder on national free education, Haldane on devolution, and Maxse on the need for better government intelligence departments. In 1904–1905 a new set of themes was introduced: the monarchy, national ethics, the race question, militarism, local government, the role of the navy, and emigration. And so it went for five years, by which time there were twenty-five members. In 1908 all the themes had been discussed, and the group dissolved.

Commentators sometimes suggest that the Co-Efficients failed to achieve any lasting organization or influence. This misses the point. The group was intended to be a dining and talking club, at which men with a range of political perspectives could exchange ideas on large problems of the day. There was never any attempt to produce agreed policies, publish a journal, form a pressure group, or establish an office to coordinate activities. Had the Co-Efficients wanted these things, they could have harnessed resources for the task. Many of them belonged to such organizations as the Victoria League, the Round Table, and the Tariff Reform League, which did have an establishment. The Co-Efficients met to exchange ideas, not to create more office work for men who already were busy.

The normal personal factors took their toll of the early members. In 1903, Sadler left to take a chair of education at the University of Manchester; Clinton Dawkins, who was massively experienced in imperial government and financial matters, died in 1905; Haldane and Grey became cabinet ministers in 1906. H. G. Wells was committing adultery with Pember Reeves's daughter and was sharpening his pen to parody the group, particularly the Webbs, in the *New Machiavelli* (1910).

Apart from the Co-Efficients, Mackinder enjoyed a range of social occasions. There were dinner parties at the houses of the Webbs and the Lyttletons and with the circle of young imperialists who surrounded Lord Milner. One dinner party at the Webbs', early in 1905, was particularly distinguished, and more than one participant wrote of the event in their diary. Beatrice Webb listed the diners: the Russells, Granville Barkers, Oliver Lodge, Mackinder, Lionel Pillimore, a Mr. Wernher, and the prime minister, Arthur Balfour. Mackinder and Wernher "chummed up," but Mrs. Webb sensed that Balfour thought them to be "mere philistine materialist administrators." Bertrand Russell also wrote down his thoughts. Wernher was described as "the chief of all the South African millionaires," who had a strong German accent that Russell thought characteristic of "all the finest types of British imperialists." And Russell went on: "Poor Mackinder made a beeline for Balfour, but got landed with me, much to my amusement. It was a sore trial to his politeness, from which he extracted himself indifferently."[20]

Most people who met Mackinder regarded him as an extremely able man. However, the established elite also regarded him as an uncouth, provincial outsider. He was "coarse grained" to Beatrice Webb, "brutal" to Bertrand Russell, and apparently "a philistine" to Balfour, the leader of the Conservative party, which Mackinder had joined in 1903. Beatrice Webb thought Mackinder a coming man in politics, but clearly he was going to find it difficult to arrive on a political scene that was still controlled by such patrician families as the Russells and the Balfours. Mackinder understood this problem, but he took heart from the emergence of Lloyd George. Writing to Michael Sadler in the spring of 1903, Mackinder said, "The advance of such a man . . . shows that opinion is plastic and ready to yield to a strong lead."[21]

There were other problems. Men like Mackinder and Sadler were not really politicians. They were reformers who wished to implement farseeing policies to rectify problems. Sadler fell out of the political picture in the maneuvering that sur-

20. Drake and Cole, eds., *Our Partnership,* pp. 300–301; *The Autobiography of Bertrand Russell,* pp. 267–69; Russell to Lucy Martin Donnelly, Feb. 8, 1905. Mackinder is described by Russell in his autobiography as "the head Beast of the School of Economics." In the 1950s Russell predicted that the next world war would be between Russia and China. He made no mention of Mackinder and his ideas on the subject.
21. Ms. Eng. Misc. C.552, Bodleian.

rounded the 1902 Education Act. He was pushed out of the intelligence section of the Board of Education by a man he had originally appointed — Robert Morant. The latter simply got on with the job of drafting an education bill that tried to reconcile all the factions involved. For Morant the bill was a political exercise.[22] Sadler had built up a vast amount of information in his intelligence bureau, information he was using as the basis of a bill that would have given the country a rational education system over the long term. The 1902 act was generally admitted to be ramshackle, but it was the best bill that could be got through Parliament. Sadler resigned in 1903, when the work of his intelligence bureau was curtailed. If educational legislation came down to simply appeasing political pressure groups, there was not much point in having a research unit that produced long-term plans. (Interestingly Britain still does not have a "national" system of secondary education, and the whole question lies in the midst of the political arena.) The fact that Sadler received no political support was indicative of problems faced in the modernization process. Sadler had a better knowledge of educational systems than anyone else in Britain. He had the intellectual ability to adapt his materials to British needs, but he was rejected by an alliance of interest groups of the type that all traditional societies possess.

Early in 1908, Amery and Lord Milner persuaded Mackinder to give up the directorship of LSE and devote himself to the cause of imperial unity. When Beatrice Webb learned the news, she wrote in her diary: "He has been the best of colleagues, during these four years, and has improved the internal organization and the external position of the School. In-

22. Birmingham University JC/18/16/16. Joseph Chamberlain went on to describe the role of Morant in the 1902 Education Act: "There was suddenly revealed a *deus ex machina* in the person of Robert Morant; a strange, enthusiastic, indomitable and hyper-industrious creature, a master of ingenuity, full of ideas, and a daring appropriator of all the technical information that lay about in the Education Office. Somehow or other Morant could adapt himself to any environment; he could capture the attention of any individual, and he had an honest and fervid interest in any subject to which he addressed himself."

deed so competent a Director has he been, that he has virtually run the whole business; Sidney trusting his initiative and executive capacity." A little later, when Mackinder's successor, Pember Reeves, yet another Co-Efficient and imperialist, took over the directorship, Webb added, "Reeves has taken hold of the School and promises to be a more staid and conscientious administrator than Mackinder, though often not so brilliant."[23]

23. Diary of Beatrice Webb, 1908, BLPES; Beatrice Webb to Mary Playne, September, 1908, in Norman Mackenzie, *Letters of Sidney and Beatrice Webb*, vol. 2, p. 316.

CHAPTER

9

Imperial Unity, Politics, and Industry

Early in 1908, Lord Milner found a means to buy Mackinder out of the directorship of LSE so that he could devote his time to the cause of imperial unity. Mackinder was to receive £850 per annum for four years from the Milner source and, in addition, would retain his readership in geography at London University. The intention was that Mackinder should lecture on imperial unity, find a parliamentary seat to contest, and get a foothold in the business world.[1]

Mackinder's interest in politics was of long standing. He had debated political topics at Epsom College and at the Oxford Union. He worked for Colonel Henry Eyre's election at Gainsborough in 1886, and he had contested the 1900 general election at Warwick and Leamington as a Liberal Imperialist, or "Limp."

In 1900 the Conservatives at Warwick and Leamington were hoping that the sitting member, Alfred Lyttleton, would be returned unopposed. They were not pleased when the Liberals produced Mackinder at the last moment. "Mr. H. J. Mackinder leads the Forlorn Hope" was the Leamington *Courier*'s headline. Forlorn hope or not, Mackinder attacked the government as bunglers over the Boer War and the next week the *Courier* described him as "a veritable iconoclast." The paper disliked

1. Mackinder to Amery, May 22, 1908 (copy), Milner Papers, 193, Bodleian. This letter is reproduced in B. W. Blouet, *Sir Halford Mackinder, 1861–1947: Some New Perspectives*, pp. 29–30. J. E. Kendle, *The Colonial and Imperial Conferences, 1887–1911*, p. 127, lists the Duke of Westminster, General Craig Sellar, Violet Markham, C. S. Goldman, and Leo Amery as the sources of the money. According to his own account, Mackinder did not know where Milner got the funds. Milner had similar arrangements with Amery and Steel-Maitland.

his radical politics and advocacy of old-age pensions, temper-
ance, improved educational facilities, and better housing for
the working population. In addition, Mackinder wanted army
reform and the abolition of social influence in the military.
Above all, he advocated imperial unity. He told the electorate
that he "believed in the Empire," and in a significant passage,
which foreshadowed the Pivot paper of 1904, he claimed:

> Little England, however true to herself, would soon be less safe
> when confronted by the military powers, the rapidly develop-
> ing resources of whose vast territories would presently enable
> them to build great fleets. No other course is open to us than
> to bind Britain and her Colonies into a league of democracies,
> defended by a united navy and an efficient army.[2]

It rapidly became clear that radical Mackinder was not "a
forlorn hope," and the safe Conservative seat was given a fright,
particularly as Lyttleton was away in South Africa on government business. When Joseph Chamberlain came down from
Birmingham to speak on behalf of Lyttleton, he obviously had
been briefed about the impression Mackinder was making, and
a considerable part of his speech was a carefully prepared attack on Mackinder's position. The attack was not vitriolic;
it was almost kindly, but it was effective. The only quarrel
Chamberlain had with Mackinder, the imperialist, was that
he was attacking another imperialist—Lyttleton. Mackinder's
ideas were fine, but they were not Liberal party policy, and if
he got into Parliament, he would have to conform to the official party line.[3] At the polls Lyttleton received 2,785 votes,
and Mackinder 1,954. He had not expected to win; rather, he
regarded the Warwick and Leamington venture as valuable experience.

In 1902, Mackinder talked with a Scottish constituency
about becoming a Parliamentary candidate, but nothing came
of this.[4] In 1903 he decided, like Hewins, to join the tariff-

2. *Leamington Spa Courier,* September 22, September 29, 1900; campaign speech quoted by Mackinder in a letter to *The Times,* October 22, 1903.
3. *Leamington Spa Courier,* October 6, 1900.
4. L. M. Cantor, "Halford Mackinder: His Contribution to Geography

reform group. In 1900 he spoke of a league of democracies with a common defense policy. In the spring of 1903 he went one step further and advocated a customs union with goods from the empire entering Britain on preferential terms, and imperial markets being shielded from foreign competition by tariffs. At this time he left the Liberal Party and joined the Unionists.

The changing of party allegiance did not involve a large philosophical shift for Mackinder. As a Liberal he belonged to the imperial wing of the party.[5] When he joined the Unionists he became a part of the imperial unity group. By and large the latter group was strongly attached to social reform; thus Mackinder did not abandon his desire to see better living conditions for working people in Britain. The major policy shift involved dropping the idea of free trade and replacing it with an imperial customs union. In the politics of his day, this appeared to be a major conversion. Viewed in the light of Mackinder's lectures on trade routes, delivered to the Institute of Bankers in 1899, however, the philosophical shift was limited. In the lectures he argued that Germany was building up a dominance of continental trade, to the disadvantage of Britain, on the basis of economic integration resulting from the development of railroads and favorable tariffs.

In the fall of 1903, Mackinder had written "The Geographical Pivot of History," which he delivered at the RGS early in 1904. The paper predicted the probable rise of a great Eurasian land power. The essay can be seen as a carefully thought out underpinning for the imperial unity movement, because if a dominant continental power were to emerge, then binding

---

and Education," p. 27. Mackinder made an entry in his diary, "Ludicrous position—in debt and coquetting with Parliament. But I mean to pull it off." During his research, Cantor was shown portions of a Mackinder diary by surviving family. I have not been able to trace the diary because the family members are now dead. Julian Amery, *Joseph Chamberlain and the Tariff Reform Campaign, The Life of Joseph Chamberlain, 1901–1903*, vol. 5, p. 306, quotes a document stating that Mackinder was the "Liberal Imperialist Candidate for the College Division of Glasgow." Had Mackinder contested the College Division in 1906, as a Limp, he might well have won the seat, which swung from Conservative to Liberal in that election.
5. H. C. G. Matthew, *The Liberal Imperialists: The Ideas and Politics of a Post-Gladstonian Elite.*

Britain and her colonies into a league of democracies would be necessary if Britain were to remain a major force in world affairs.

Although the tariff-reform ideas fitted well into Mackinder's view of imperial development, the issue could not have come at a worse time for him politically. By accepting the tariff-reform position, he was forced to leave the Liberal party and missed the chance of being considered for office in the administration that party formed in 1906. Mackinder did not gain political advancement from his alliance with the tariff reformers. In the weeks after Chamberlain's May 15 call for tariffs a nonparty body—the Stafford House Group—emerged to promote tariff reform. Amery and Mackinder were prominent in this effort, and when the Tariff Reform League was founded, on July 21, 1903, they represented Stafford House. When a secretary for the new League was sought, Amery and Steel-Maitland pushed for Mackinder, who had ideas on how to organize the League. In the end, however, the selection committee decided upon a lawyer with administrative skills more mundane than Mackinder's. The disappointment was probably a blessing, for Mackinder could not have taken up the directorship of the London School of Economics had he become secretary of the League.[6]

Conversely, having accepted the directorship, later in 1903, he could hardly contest the 1906 election without leaving LSE. Amery corresponded with Joseph Chamberlain about finding Mackinder a seat as a conservative candidate, but nothing came of the matter.[7] Mackinder was active in the campaign, speaking on behalf of Amery, Steel-Maitland, and Lyttleton in the Midlands. There was some fuss at Leamington, where Mackinder tried to speak in support of Lyttleton, the candidate he had previously stood against, and at one meeting he was refused a hearing.

Just prior to the 1906 election, Mackinder wrote *Money-*

6. Porter, *The Unionist Tariff Reformers, 1903–1914*, pp. 234–36; D. P. Cook, *Benjamin Kidd: Portrait of a Social Darwinist*, p. 224. Cook's study treats Kidd's commitment to tariff reform.
7. Joseph Chamberlain to Amery, November 24, 1905, J. C. 21/1/14, Birmingham University.

Power and Man-Power: The Underlying Principles Rather Than the Statistics of Tariff Reform. The pamphlet set out Mackinder's view of Britain's position in the world and what might be done to sustain her place among the great states.

> Let us regard Power, Trade, Wages, and Labour as forming a circle. . . . Now Power shelters Trade, Trade supplies Wages, Wages maintain Labour, and Labour is the source of power. Much Power is needed to shelter a great Trade. A great Trade can alone supply much wages and support a great and efficient population. A great and efficient production is the only firm source of great Power. . . . The great states of the world have lately increased . . . in Power. Therefore, if Britain is to remain among them, a broader base of Trade, Wages, and Labour must be found for her. . . .
>
> The object of the policy of Tariff Reform is to promote British trade in the ways best calculated to increase wages and population, and so to supply the needful broader base for our Power.[8]

The pamphlet was marked by some writing that was laborious by Mackinder's standards, but the essay was widely circulated. It reflected the view of Amery and Hewins that economic power was the basis of imperial power, and it became a source of ideas for the imperial unity group. Lord Milner made use of it when constructing speeches on imperial themes.

In the summer of 1908, Mackinder gave up the directorship of LSE and his seat on the senate of London University. He sailed to Canada on a battleship to begin a tour that lasted for several months. If the idea of imperial unity was to make progress, then agreements had to be reached with Canada, as it was the most important of the self-governing dominions. In addition it was under economic pressure from the United States, and closer ties with Britain might result in advantages. Mackinder met many Canadians and gave talks across the

8. H. J. Mackinder, *Money-Power and Man-Power: The Underlying Principles Rather Than the Statistics of Tariff Reform.* A fragment of diary accompanying the copy of the pamphlet in the Mackinder Papers, School of Geography, Oxford, tells something of Mackinder's speaking engagements in the 1906 election.

land, including addresses to the influential Canada Club, in Winnipeg, on "British Sea Power" (September 10, 1908), and at the Ottawa branch, on "Canada and Empire Problems" (September 24, 1908). In the sea-power speech he developed a theme used repeatedly by Milner and others when speaking of the importance of the navy: the Royal Navy represented a credit at the bank of world power, and Britain did not have to fight in many situations because it had credit at that bank. At the end of the tour, Mackinder consulted in Montreal with Milner, who was on a visit to deliver speeches on imperial unity and defense.[9]

On his return to England, in late 1908, Mackinder gave three talks to the Compatriots in which he outlined the problems of a relatively undeveloped region, Canada, alongside an industrially advanced country, the United States. He was of the view that Canada was not "a field, a forest, and a quarry" for the United States and that the dominion could be strengthened by the development of closer economic links with Britain.[10] Throughout his political career, Mackinder spoke knowledgeably on Canadian affairs, and his experience of the country was particularly valuable when he came to chair the Imperial Shipping Committee.

Gaining a Seat in Parliament

On returning to Britain, Mackinder searched for a parliamentary seat. In February 1909 he was adopted as the Conservative and Unionist candidate for Hawick Burghs, where a by-election was to be fought. In 1906 the seat had been contested for the Conservatives by Sir Arthur Conan Doyle, who lost by a large margin to the Liberal candidate. Mackinder did better in 1909 but still lost by more than five hundred votes. The Liberal, Sir John Barran, polled 3,028 votes, and Mackinder received 2,508. Necessarily the campaign was based upon big

9. Diary, September 19, 1908, Milner 271, Bodleian.
10. *The Times* December 1, December 8, December 15, 1908. The Compatriots was established to promote the "ideal of the wider patriotism of the commonwealth" (L. S. Amery, *My Political Life*, p. 265).

meetings and set-piece speeches, for there was no time to build a local organization—as Hewins noticed when he came up to speak on Mackinder's behalf.[11]

The loss was expected, and it did him no harm. On the contrary, Scottish Conservatives and Liberal Unionists were impressed with Mackinder's ability as a public speaker and the force of his imperial message. He was invited to address political meetings and was sought after as a prospective candidate in Glasgow. The St. Rollox, Camlachie, and Tradeston constituencies expressed interest, and in June it was announced that he would fight Camlachie in the next election.[12] He might well have found a safer seat, but he wanted a working man's constituency, where he felt he could address the problems of the time.

Camlachie was not an easy constituency for a Conservative and Unionist. The voters were mostly working men engaged in shipbuilding and other heavy industry vulnerable to recession. There was an Orange element and a strong Roman Catholic presence. The constituency had a good following for both the Liberal and Labour parties; this was an advantage to Mackinder as long as they ran candidates and divided the left-of-center votes.

Mackinder's speeches in 1909 and 1910 were reported in the *Glasgow Herald*. The paper thought him one of the strong-

11. Feb. 16, 1909, MS 196, H. P. "I formed a very poor impression of Mackinder's chances. Like Amery at Wolverhampton he is depending too much on big meetings. His address is very poor stuff." Hewins found out how difficult it was to win a parliamentary seat when he lost the Shipley division badly in January 1910.
12. *Glasgow Herald*, June 8, 1909. Mackinder's selection at Camlachie was complicated. The Conservatives and the Liberal Unionists in the constituency had agreed to run joint candidates. In the 1906 general election the two factions had supported A. Cross, the choice of the Liberal Unionists. Cross won the election and then tended to support the Liberal government rather than the Conservative and Unionist opposition. The Conservatives in Camlachie were unhappy with this and told the Liberal Unionists to find another candidate for the 1910 election. Mackinder was that candidate, and at the election he defeated Cross, who stood as a Liberal. Minute Book No. 3, West of Scotland Liberal Unionist Association, The Scottish Conservative Central Office, Edinburgh. I am indebted to Ann Hay for help with Conservative party records.

est Conservative and Unionist candidates in Scotland and approved of the "clarity of his imperial vision." Mackinder's campaign speeches outlined the policy he wanted for Britain in response to the rising great states in Europe, to which he had referred in the 1904 Pivot paper. There was no mention of acquiring more territory; the Empire was in a phase of consolidation. He wanted to see "a group of . . . nations, the Britains, with one fleet on the ocean . . . and one foreign policy" (*Glasgow Herald*, April 6, 1909). The need for a strong navy was stressed continually as the means of maintaining Britain's place in world affairs. There was no mention of territorial involvement in Eurasia as a means of curtailing the possible growth of a pivot power.

Mackinder advocated the abandonment of free trade. "Britain had been driven from near markets to far as Europe and America had built up industries behind tariff walls" (April 6). A little later, on April 27, he commented, "The nations of the world declared by tariffs that they meant to maintain nationalities, and however grand our ideals might be we were powerless to play our game [free trade] on the face of the earth if people would not play with us." Mackinder thought the country had slipped to a bad third in the league of industrial nations. British shipping and banking were predominant, but in steelmaking, for example, the United States and Germany had larger and more efficient manufacturing plants. Britain had an investment problem. Too much capital was being exported, and emigration represented a loss of trained manpower. Germany had a very low rate of emigration, and Belgium, a country with a population density higher than Britain's, had low emigration because she created more jobs by investment.[13]

Mackinder argued that tariff reform could correct the problems. In a long speech at his formal adoption as candidate for

13. Andrew Gamble, *Britain in Decline*, pp. 52–63, is an excellent summary of the tariff-reform perspective. Gamble, like Mackinder, is of the opinion that Britain adhered to free trade "long after the conditions which had originally recommended it had disappeared" (p. 58). Gamble also makes the interesting point that in contesting German claims to continental dominance in two world wars, Britain weakened herself and conceded leadership of the oceanic realm to the United States.

Camlachie, reported in the *Glasgow Herald* on December 21, 1909, he described how the new tariffs would work. He saw them as encouraging more investment and more employment in Britain. Tariffs could be used to negotiate the mutual lowering of duties with other countries, dues could be high on luxurious imports, and the foreigner exporting to Britain could be made to pay a part of the taxes. Colonies would be given preferential treatment. The speech had almost a post–World War II tone, and some of the arguments eventually were used to promote the European Common Market.

It is sometimes suggested that the tariff reformers failed.[14] In the sense that they did not get their ideas implemented before 1914, this is true. But the idea of imperial preference was embraced in the Ottawa Agreement of 1932, and Britain has accepted the general proposition that investment, industry, and employment could be strengthened behind protective tariffs. The country lies within a common external tariff, but it is a tariff established by the European Common Market. Whether or not it would have been possible to create a common market of the British Empire and what impact this might have had upon economic development in the Third World are hypothetical questions.

Fundamentally, Mackinder saw two major problems in the first decade of the twentieth century: how was the position of the "Britains" in international affairs to be maintained, and how was the relative industrial decline of the United Kingdom to be reversed? As we have seen, tariff reformers felt that these two elements were a part of the same picture, and as reported in the *Glasgow Herald* (June 6, 1910), Mackinder thought that:

> We had gone on blindly, saying that we were the greatest industrial nation in the world. But the situation had changed, and he

14. Alan Sykes, *Tariff Reform in British Politics, 1903–1913*, pp. 285–94. Sykes points out that "all the leaders of the Conservative party between Balfour's resignation and the Second World War came originally from the tariff reform wing of the party, and moreover from the extreme end of that wing." This could be an indication that members of the tariff-reform group, such as Mackinder, generally held ideas that were ahead of their time. See also D. Porter, "The Unionist Tariff Reformers."

ventured to say that now was one of the last, if not the last time we had to accommodate ourselves to the new conditions. Why did he put it this sternly? For the reasons that we had held our own during the past 30 years by brute force . . . driven to trust to the strength of our right arm, our Navy, and had used the Navy mercilessly, though silently, because it was all powerful and the foreigners knew it.

A theme Mackinder constantly spoke of was the danger of the German naval buildup, which he saw as a threat to Britain's position in the world. Although he traveled frequently to Germany, and enjoyed his visits, his views contributed to anti-German feeling. It is probable that when he warned of the rise of a great land power in 1904, he had Germany in mind.

At one political meeting in 1910, when he was speaking of the German naval threat, a heckler shouted:

"Declare war against them, then."
"No," replied Mackinder, "I do not want to declare war."
"You preach it," came the response.[15]

Even though Mackinder went on to argue that a strong navy had averted war, he had lost the exchange with the heckler. He was not a good rough-and-tumble politician, and he never added to his skill as an orator the ability to deliver the sharp reply.

The information presented so far characterizes Mackinder as being concerned with imperial defense and the competitive position of British big business. However, if we analyze the full range of his activities, we uncover a more complex picture. Mackinder was active in the affairs of the Victoria League, an organization established in 1901 to promote friendship between parts of the Empire by means of educational programs. From 1903 he was a leading figure on the Colonial Office visual-instruction committee. In fact he persuaded the office to create mechanisms whereby children in different parts of the Empire could learn about other British possessions. Mackinder wrote scripts to accompany carefully prepared slide sets. In 1907, Hugh Fisher was sent around the Empire, under

15. *Glasgow Herald*, January 5, 1910.

instructions from Mackinder, to record scenes and life styles across the culturally diverse British possessions.[16]

Mackinder's views on the relationship between cultures within the Empire during this period were set out in a piece entitled *On Thinking Imperially*. He took the view that it was wrong to see the Empire as a system in which Britain was a manufacturing center and the overseas territories were producers of food and raw materials. Industrial capacity had to be more widespread. The educational system in Britain should not teach history in an insular way and should imbue "in the children of an Imperial race . . . a freedom from contempt of other races." The English had to stop thinking, for example, of Moslems as pagans, and the Empire must maintain its "provincial nationalities" and cultivate the idea of equality between those nationalities.[17]

It is difficult to categorize Mackinder's brand of imperialism. He believed in a strong imperial defense policy as part of the effort needed to keep Britain in the front rank of the world powers. At the same time he subscribed to the idea that the Empire was to become a Commonwealth of Nations consisting of many cultures and peoples who were associated economically and politically. As we now struggle with the problems of the North-South economic dichotomy in world affairs, it is time to look back at the ideas of men like Milner and Mackinder, and reassess them in the experience of the twentieth century. They had a vision of associations between culturally and economically diverse regions that makes the European Economic Community appear a pedestrian bureaucracy.

Mackinder's interest in imperial causes did not lessen his concern for social problems of the type he spoke of in his Leamington campaign of 1900. In 1910 he was still a tariff reformer "because he wanted social reform" (*Glasgow Herald*, March 23, 1910). His choice of a working-class constituency in Glasgow was deliberate, and he referred to the "unwholesome conditions of life in Camlachie" (October 16, 1909). He could climb onto a platform with Beatrice Webb, in November 1910, as she

16. C.O. 885/17, 21, 19, PRO. The pictures are in the possession of the Royal Commonwealth Society.
17. H. J. Mackinder, "On Thinking Imperially," in M. E. Sadler, *Lectures on Empire*, pp. 34–42 (quote from pp. 36–37).

campaigned for her minority report on Poor Law Reform, support her ideas, and claim it was wrong to denounce the report as "Socialistic," which Asquith and Balfour had done. This is not to say he was a socialist; on the contrary, he frequently denounced the idea of socialism. What he meant by socialism was the growth of bureaucracy and being "ruled by officials" (October 28, 1910). He fought against the Lloyd George budget of 1910, on the grounds that it taxed savings and set up a much more complex government machine.

It is easy to mistake Mackinder's position in relation to great power competition. He constantly talked of the emergence of what we would call a superpower in Eurasia. He advocated a strong imperial defense policy to counter the danger, and it is easy to believe, along with Beatrice Webb, that he thought the world should be divided into great states maintaining some type of military balance between themselves. But this view oversimplifies his position. Mackinder believed that a superpower was coming and that if the "Britains" wanted to protect themselves effectively, they should create a "league of democracies." At the same time he sought to preserve provincial identity within the league and guard against too great a concentration of power in the machinery of government. The fear of too much bureaucracy was probably a product of the nineteenth-century attitudes Mackinder had encountered during his youth, and the desire to protect provincial identity owed something to the fact that Mackinder, and many of his associates, had provincial origins.

To the *Glasgow Herald,* and to the voters of Camlachie, it was "not a question of present importance whether all his views are realizable." They liked the look of Mackinder's abilities sufficiently well for him to win the election of January 1910. Mackinder (Conservative and Unionist), received 3,227 votes. Cross (Liberal) got 2,793 and H. S. Kessack (Labour) 2,443. The majority was a reasonable margin, but clearly he owed victory, in part, to the strength of the Labour following, which had cut into Liberal support. In the Glasgow constituency of St. Rollox, which Mackinder could have fought, there was no Labour candidate and the Liberal majority was more than 3,000.

Mackinder made his maiden parliamentary speech in the debate on the King's speech and was well received by such tariff reformers as Austen Chamberlain.[18] He set out many of his ideas in a broad context and argued that questions relating to the export of capital and people, the decline in industrial capacity and in Britain's position in the world were all interlinked. Alfred Mond, a prominent member of the Free Trade League, denounced these ideas as "wild and ridiculous." In general Mackinder was not well received in the House of Commons as a speaker: his university lecturing style was resented, however knowledgeable the exposition.

By the end of 1910 he was back in Glasgow to fight the second general election of the year, which had been precipitated by Lloyd George's budget. Mackinder received more votes than previously (3,479), but a new candidate, J. M. Hogge, produced a big increase in the Liberal vote (3,453), and the Labour vote slumped to 1,539. Mackinder's majority was just 26 votes! There was a fourth candidate, a Suffragist, W. J. Mirrlees, one of only two standing in the country, who got 36 votes. Had Mr. Mirrlees attracted just a few more votes, he could have cost Mackinder the election. Mackinder was prepared to support a referendum on the question of votes for women, but his opinion was summed up in the phrase that women "wished for the privilege of one sex and the power of the other."[19]

Before 1914, Mackinder's contributions to debate in the House of Commons were sporadic because he was busy trying to make a living. But he built up his knowledge of colonial and imperial affairs, merchant shipping, and Scottish problems, and he spoke in debates on these matters. In April 1913 he introduced a motion proposing that "there should be established in London a National Theatre, to be vested in trustees and assisted by the State, for the performance of the plays of Shakespeare and other dramas of recognized merit." Nothing came of this at the time because leading figures in the theater world were unsure of the type of national theater they wanted.[20]

18. *Parliamentary Debates*, Fifth Series, vol. 19, February 21, 1910;
   Austen Chamberlain, *Politics from the Inside*, p. 204.
19. *Glasgow Herald*, January 21, 1911.
20. Ms 656, Coll. Misc. 482, BLPES, contains Mackinder materials relating to the National Theatre project.

World War I brought Mackinder fully into public life. On August 7, 1914, Lord Kitchener issued his first appeal for volunteers to enter the British army. On August 12 the secretary of war called upon the political parties of Scotland to help enroll volunteers. Mackinder took the leading role in this effort. In a few days the rooms of political parties throughout Scotland were opened as recruiting centers. The centers enlisted 213,751 volunteers prior to the Military Services Act, which introduced conscription on March 2, 1916.[21] Mackinder devoted himself to the task of attracting volunteers and used his power of oratory to encourage men to enlist. Haldane, who had known Halford for many years in connection with LSE, suggested to Kitchener that he use Mackinder to help organize recruiting in Scotland.[22]

Mackinder's contributions to public life in the early years of the war had a populist tone. He helped with recruiting, suggested a national savings scheme which benefited the small saver, and spoke in the House about the danger of the coalition government's getting out of touch with the majority of M.P.s and the country. But he did not penetrate the inner ranks of government and find a job that could fully employ his administrative and intellectual powers. Mackinder had a patron in the War Cabinet, Lord Milner, but he lacked a strong political base and was not given office. The war gave Mackinder opportunities to display talents, but because the usual democratic processes were suspended, it was not easy to advance. The entrenched elements were able to ignore the able M.P. from Scotland who tended, in the establishment view, to be too supportive of such popular causes as the 1917 rent strike in Glasgow.[23]

21. Minute Book No. 1, Central Council, Scottish Unionist Association, p. 43, The Scottish Conservative Central Office, Edinburgh. Kitchener's call for volunteers raised nearly two and a half million men for the army (Philip Magnus, *Kitchener: Portrait of an Imperialist*, p. 345).
22. W. H. Parker, *Mackinder: Geography as an Aid to Statecraft*, p. 43.
23. For an example of Mackinder's wartime populist tone in Parliament, see the report of his speech November 10, 1915, *Parliamentary Debates*, Fifth Series, vol. 75, in which he argues that the coalition government is growing "out of touch with the nation" and

The Electro Bleach Company

Members of Parliament began to receive a small salary in 1911, but Mackinder still did not have the financial resources to sustain the activities of an M.P. The few hundred pounds a year he received for teaching at London University, and the income from books, was insufficient, particularly when the Milner monies expired in 1912. He had hoped that he would get directorships in industry. When offers did not come, he started companies of his own.

In 1911 he became involved in a machine-tool enterprise and began, along with James Swinburne, FRS, (1858–1958), to develop a scheme to get into the Cheshire salt industry and manufacture a range of products, using as a stock the Middlewich brine deposit. In 1913 a firm called Electrolytic Alkali Company was closed, and its assets, consisting of plant and land underlain by salt deposits, were sold to the Electro Bleach Company.[24] Electro Bleach consisted of Mackinder, Swinburne, F. J. Dundas, a colliery director, and E. G. Cubitt, who had an interest in a fertilizer company. Electro Bleach was to use the Hargreaves Bird process to manufacture a range of products, particularly bleaches, for the textile and papermaking industries. The Hargreaves Bird process was reputed to be efficient and, because the plant sat on top of a brine deposit, there were hopes that the venture would be profitable.

The assets of Electro Bleach were valued at £281,326, but the directors did not have that much capital. On March 14, 1914, Electro Bleach successfully offered £160,000 worth of stock to the public. Unfortunately the company had not started to manufacture by the time the 1914–18 war began, and in a time of financial uncertainty, many of those who had wanted stock were slow to deliver funds.

In spite of the difficulties, production was started, and in 1915 a modest profit made. Activity increased when the gas

the need of the government is to "place before the great masses of the people some idea of the scope of its policy."

24. Information on Electro Bleach has been derived from the *Stock Exchange Yearbook* and company reports published in *The Times*.

war started on the Western front and the Ministry of Munitions ordered large quantities of chlorine. In the circumstances the company might have been expected to make good financial returns, but in 1916 an excess profits tax was introduced. The tax worked against firms that had made losses before the war, or had not traded. If high profits had been made before the war, then high profits in the war years escaped the full effect of the new tax.

As production at Electro Bleach increased, a capital shortage developed and the Ministry of Munitions advanced £35,000 to finance additional bore holes. Overall the company tended to become a subsidiary of the ministry, which gave a fixed percentage return to Electro Bleach over the cost of production. The shareholders never got good dividends, and at the end of the war a telegram arrived from the ministry ordering a stop to the production of materials and a return to peacetime production. After the war, capital was needed to pay off the ministry loan and to replace worn equipment. In November 1919 the capital of the company was increased to £480,000 by the issue of new shares for which existing shareholders were invited to subscribe. In 1920 Brunner Mond and Company (later ICI) made a successful offer for the company, and Mackinder disappeared from the board. Production was stopped at Electro Bleach in 1925 as part of a rationalization process.

Although Mackinder was not made rich by the takeover, for five or six years Electro Bleach did provide him with an income that allowed him to take part more fully in parliamentary affairs. Overall he made money sufficient to pay for his public life, but he never became financially independent. He was against salaries for members of Parliament, but his own career provided a strong argument for paying M.P.s.

The Elections of 1918 and 1922

The House of Commons may not have appreciated Mackinder the lecturer, but it came to value his ability to get to the heart of a problem and suggest new policies. After the start of the war he was appointed to a number of committees and did excellent work. From 1914 to 1916 he served on a committee that looked into the question of war loans for the small investor, chaired by E. S. Montagu. Then he worked on the National War Savings Committee, where he suggested the idea of saving stamps, which encouraged millions of small savers to contribute to the financial needs of the war effort. The scheme was still in operation for World War II. He served on committees looking into the standardization of railway equipment (1918) and the regionalization of British railroads (1921). He was a member of royal commissions on income tax (1919), awards to inventors (1919), and after he had left Parliament, on food prices (1924–25).[25] For these public and Parliamentary services, he was knighted in 1920.

In the election of 1918, Mackinder retained his seat in Parliament with ease. He received more than thirteen thousand votes, nearly double the number polled by the Labour candidate.[26] The Liberal candidate received fewer than a thousand votes, which indicated that the constituency was polarizing between Labour and Conservative. In a working-class area of Glasgow, which Camlachie was, the trend would undermine Mackinder's position.

The election of 1922 was precipitated when, at a famous meeting at the Carlton Club, the Conservative M.P.s voted to cease cooperating with the Liberal leader, Lloyd George, in the formation of a government. Mackinder was against breaking up the coalition but was on the losing side in his own party.[27]

The *Glasgow Herald* expressed the view that a loss by Mackinder in the 1922 election would be "an improbable contingency" (November 4, 1922). Mackinder had built up broad support with his handling of Scottish questions in Parliament. He was chairman of the Scottish members committee,

25. P. Ford and G. Ford, *A Breviate of Parliamentary Papers, 1900–1916*, pp. 68–69; *Breviate, 1917–1939*, pp. 258–59.
26. *Glasgow Herald*, December 30, 1918.
27. Michael Kennear, *The Fall of Lloyd George*, p. 234. By the end of his political career, Mackinder was regarded, at least by the *Glasgow Herald*, as a moderate. Interestingly, many of the people Mackinder had been associated with—Amery, Steel-Maitland, Marriott, and Oman—voted against the coalition. See also, K. O. Morgan, *Consensus and Disunity: The Lloyd George Coalition Government, 1918–1929*, p. 354.

and his political speeches during and after the war were marked by efforts to reconcile conflicting interests.

On the other hand, he had lost support on three sensitive issues. As a result of his involvement in the South Russia episode, left-wing elements worked hard against him. His statement early in 1922 about the dangers of teaching propaganda in schools had brought criticism from some churchmen. The third issue that cost him dearly was Ireland. The treaty creating the Irish Free State was signed on December 6, 1921, by Lloyd George and Irish representatives. Early in 1922 Mackinder urged that the experiment be given a chance and expressed the view that three million people in the south of Ireland should not be allowed "to hold up the destinies of humanity." He felt that the independence of Ireland should be accepted, for it gave moral strength to the British Empire (*Glasgow Herald*, January 20, 1922). Although this was a commonsense view, it was likely to be costly in a constituency where many voters had traditionally supported the idea of union with Ireland. In retrospect, it is hardly surprising that the Labour candidate, the Reverend Campbell Stephen, a teacher in the employ of the Glasgow Education Authority, won fairly easily, with 15,181 votes to Mackinder's 11,459.

It is probably unrealistic to think that Mackinder would have been given a Cabinet post in the 1906 Liberal government had he remained in that party. But in 1922 he was poised to take office. At age sixty-two, Sir Halford was a senior M.P., a known public figure with wide experience and a moderate reputation. L. S. Amery was pushing for him to be offered the Ministry of Education or the Board of Trade.[28] The Conservative party won the 1922 election. When forming an administration, however, it was generally recognized to be short of experienced men of strong intellect. Had Mackinder retained the Camlachie seat he would have been a powerful contender for office. He lost the election and the chance of becoming a minister.

There is evidence that Mackinder was tired of party poli-

28. John Barnes and David Nicholson, eds., *The Leo Amery Diaries*, vol. 1: 1896–1929, pp. 300–301.

tics. He did not campaign as hard in Camlachie as he had done previously. In fact, Clydeside was one of the few areas of the country that swung toward Labour in the 1922 election. Once he lost the seat, he declined the Camlachie offer to remain as prospective Conservative and Unionist candidate. Not surprisingly, given his senior status and experience, he was offered the chance to fight safer seats, including Glasgow Central, and Selkirk and Peebles. All offers were turned aside. Since 1920, Sir Halford had served as chairman of the Imperial Shipping Committee. The committee work was absorbing; in it he saw the opportunity to promote, in practical ways, economic cooperation within the Empire.

10

Democratic Ideals and Reality

The 1914–18 war was the first in which academics, as a group, had an impact on policy making. In Britain academics wrote the Admiralty handbooks on foreign areas, and in America geographers such as Bowman, Jefferson, Johnson, and Semple developed material for use in the boundary-making process. In France extensive studies for a future peace were conducted by an important group of scholars chaired by E. Lavisse and Vidal de la Blache. Many of the British academic state and boundary makers formed a pressure group that worked against the concept of negotiated peace with Germany.

The Boundaries of Europe

Among the British geographers the imaging of new boundaries started early. On December 7, 1914, L. W. Lyde, University College, London, delivered a paper at the Royal Geographical Society on the political frontiers of Europe. Lyde tried to establish the theoretical bases of boundary making, but he was not received with much enthusiasm. In discussion Spenser Wilkinson expressed the view that Lyde's exercise was premature, and Mackinder added:

> Some people . . . think you will at the end of the war be able to set up a new Europe in accordance with scientific ideals. I am not so sure. I think you will find that the old idea of the balance of power will assert itself again in any congress in Europe. . . . Germany will be a nation of 70 million in the centre of Europe. . . . [I]t will still be so strong a power that I question whether there will be very much of ideal map making. If you conquer that power, the object will be to clip its wings for the future.

He went on to outline some of the problems of creating new states. For example, "the region inhabited by Poles lies wholly inland," but if the new Poland received German land to provide access to the sea, then the boundary makers would have "set up a new Alsace-Lorraine."[1]

From this RGS meeting, the evolution of Mackinder's ideas on the postwar boundaries of Europe can be traced in a series of articles and statements that culminated in his book *Democratic Ideals and Reality*, published in 1919 in Britain and the United States. Early in 1915, suggesting that the day of "great Kaiserdoms" had passed, Mackinder looked forward to the emergence of a constellation of minor nationalities in southeast Europe. Perhaps, he suggested, the new nations, which would include Hungary, Roumania, Serbia, and Bulgaria, could be federated to provide strength.[2] He was convinced that the war had been precipitated, in part, by the development of a highly centralized state in Germany. He saw overcentralization of power as a problem and, on these grounds, advocated the creation of smaller, viable states. It was an extension of his argument on the importance of retaining provincial identities.

In 1916, Mackinder devoted an increasing amount of time to the problem of state and boundary making. He helped found the Serbian Society, an organization that promoted the idea of the state that eventually became Yugoslavia.[3] In an article in the *Glasgow Herald* (September 30, 1916) he saw Serbia as incorporating Bosnia, Herzegovina, Slovenia, and Croatia. He suggested that the new Serbia and Roumania might form an alliance to provide regional stability.

Mackinder was not, of course, developing his ideas in isola-

1. L. W. Lyde, "Types of Political Frontiers in Europe," *Geographical Journal* 45 (1915): 126–45. See also Lyde's *Some Frontiers of Tomorrow: An Aspiration for Europe*.
2. H. J. Mackinder, *Glasgow Herald*, January 30, 1915.
3. Hugh and Christopher Seton-Watson, *The Making of a New Europe: R. W. Seton-Watson and the Last Years of Austria-Hungary*. For studies of official British attitudes toward East and Central European subject nations, see Kenneth J. Calder, *Britain and the Origins of the New Europe, 1914–1918*, and Wilfred Fest, *Peace or Partition: The Habsburg Monarchy and British Policy, 1914–1918*.

tion, and in this period he was drawn into the New Europe group. The group had a strong academic component and revolved around the historian R. W. Seton-Watson (1879–1951) and Tomáš G. Masaryk (1850–1937). Masaryk had taught at Prague University, but in December 1914 he escaped to Britain and took a post at Kings College, London University. He became president of the new state of Czechoslovakia in 1918.[4] Seton-Watson had established himself as an authority on the Balkans and Austro-Hungaria long before the war started. He wanted to turn back the Pan-German movement embodied in ideas of *Mitteleuropa* and a Berlin-to-Baghdad axis. Further he wanted to see:

> The creation of a new Europe upon a mainly racial basis—the reduction of Germany to her national boundaries, the restoration of Polish and Bohemian independence, the completion of Italian, Roumanian, Yugoslav, and Greek national unity, the ejection of the Turks from Europe.

Seton-Watson was regarded as a dangerous agitator by some in Britain. As a consequence, even though he was medically unfit for military service, he was conscripted for a time in an unsuccessful effort to curtail his activities. Mackinder, as an M.P., helped gain his release from the military. In October 1916, Seton-Watson published the first issue of a weekly entitled *The New Europe*. The aim was to create a climate in which "our statesman will win the peace" and promote "emancipation in the subject races of central and south-eastern Europe from German and Magyar control."[5]

Mackinder was a useful, but not central, member of the New Europe group. From Masaryk and others he learned a great deal about German ideas for economic control of Central Europe, which strengthened his fear of Germany's long-term role in Europe. Mackinder had commissioned and edited

4. T. G. Masaryk, *The New Europe. Nova Europa* (Prague, 1920) has maps to illustrate Masaryk's ideas on Europe.
5. R. W. Seton-Watson, *Europe in the Melting Pot*, p. x; Hugh and Christopher Seton-Watson, *The Making of a New Europe*, p. 198. *The New Europe* was published from 1916 until 1920. Seton-Watson, *The Making of a New Europe*, contains a discussion of the journal's role.

Partsch's book, *Central Europe* (1903). In this work there were hints about equilibrium being established if "the Powers of Central Europe stand shoulder to shoulder for the maintenance of the free and peaceful economic development which must reach even further and further abroad, as the increasing populations find their homes growing too narrow for them." These notions found a place in Naumann's *Mitteleuropa* (1915), which was widely reviewed in the geographical literature and contained suggestions for a Central European customs union.[6] In some quarters Naumann's work was seen as containing the basis for negotiated peace, but Tomáš Masaryk, in a series of articles on pan-Germanism published in *The New Europe,* put the matter in a quite different light. Masaryk traced the extent of German academic writing and theorizing on the economic development and control of Central Europe back to Friedrich List and his ideas for a Central European state. Probably for the first time, many in Britain became aware of the extent to which German academia was harnessed in the cause of *Real-Politik.*[7]

At the outset of war, Mackinder pondered whether or not it was possible to create new states as a part of the peace process. By 1917 he was supporting the national aspirations of the peoples of East Europe.[8] He was not motivated solely by democratic ideals. He wanted to cut Germany down to a manageable size, to curtail German economic power, and he had a general desire to see smaller states. Big organizations were

6. Josef Partsch, *Central Europe,* p. 159. In chapter 7, "Tariff Problems," Naumann set out much of the theory that lies behind the common-market idea. Of course, Naumann was advocating a Central European economic partnership centered upon Germany. This was alarming to the British tariff-reform group, who argued that, however reasonable the notion sounded, it would greatly increase German strength. The book was published in English as *Central Europe.* The far-reaching nature of Germany's economic aims is outlined in Fritz Fischer, *Germany's Aims in the First World War.*

7. Thomas G. Masaryk, "Pangermanism and the Eastern Question," *The New Europe* 1 (1916): 2–19, and "The Literature of Pangermanism," *The New Europe* 1 (1916): 57–60, 89–92, 118–24, 152–57, 247–49.

8. H. J. Mackinder, "This Unprecedented War," *Glasgow Herald,* August 4, 1917.

a threat to democracy, and he came to advocate more local administrative control in Europe, Britain, and the Empire. In 1917 he was listed as one of the collaborators on the inside cover of *The New Europe*. In May of that year, speaking at a meeting of Inter-Allied Parliamentary delegates at the Sorbonne, he advanced the idea that Britain, France, and Italy might form an economic grouping. Throughout the year, while he was involved with the south Slav issue, he kept in touch with Italian, Serbian, and Greek views on Adriatic matters and particularly Dalmatia. He wrote articles on the coming peace settlement for the *Glasgow Herald*, whose editors were always eager to have his opinions on the big questions of the day. On October 31, 1918, the paper carried Mackinder's views on the new states he wanted to see emerge:

> We are about to construct a new map of Europe upon which there will be a Poland, of some 20 million people, including a piece of Austria, a Great Bohemia of some 8 or 10 million people . . . a Hungary of some 8 or 10 million of Magyars and Jews; a Great Ukraine, mainly in Russia, but including the eastern half of Austrian Galicia; a Great Roumania containing not merely the historic principalities of Moldavia and Wallachia but also Transylvania, hitherto a part of Hungary; and a Jugo-Slavia of some 8 to 10 millions.

In addition he wanted to ensure that Germany and Austria did not form a union. Not all the ideas outlined above came to be. For example, a Great Ukraine did not emerge. Even so, Mackinder was surprised at how far the allied delegates to the peace conference did go in establishing the new states.[9]

The Publication of *Democratic Ideals and Reality*

Late in 1918, having won reelection to Parliament, Mackinder completed *Democratic Ideals and Reality*. The book was issued by Constable, a publisher with strong connections to the

9. *Glasgow Herald*, May 8, 1917 (The original text of the speech, in French, is in *The New Europe* 3 (1917): 150–53.); "End of an Empire: The Break-up of Austria-Hungary," *Glasgow Herald*, October 31,

New Europe group. Because the volume was written in haste, some of the arguments were not fully worked out, but it contained a world perspective that had evolved over two decades. The views on postwar boundaries were a product of both daily contact with the issues and many conversations with Frenchmen, Italians, and Slavs. The book supported such ideas as self-determination and the League of Nations, but it warned that, over the longer term, considerable threats would have to be overcome. Mackinder was wary of creating states that had boundaries extending beyond the ethnic group, or groups, that would form the core of a new country. He stuck to his view, for example, that the Danzig corridor would give trouble sooner or later. He even suggested that such populations as the Germans occupying territory within Poland should be resettled in order to tidy up some of the situations. In addition, because the book contained much on economic reconstruction, it became a part of the literature leading to the emergence of the common-market idea.[10]

After World War I a large number of books on the problems of the postwar world were published. *Democratic Ideals and Reality* did not stand out as an important work and was not widely noticed in journals and newspapers at the time. The reviews were mixed. *The Spectator*, as usual, found Mackinder's views interesting, pointing out that anyone who assumes that victory has made a new world "will be annoyed by Mr. Mackinder's thoughtful book." And several reviewers were irritated. Some found it fanciful, others thought it old fashioned. F. J. Teggart (Berkeley) reviewed the book for the *American Historical Review*. Teggart was an admirer of the Pivot paper but thought that there was little new in *Democratic Ideals*. Reflecting a widespread postwar euphoria, Teggart further sug-

1918. Articles by Mackinder in the *Glasgow Herald* on the boundaries of Europe include: "This Unprecedented War," August 4, 1917; "Adriatic Question," December 3, 1917; "Rome Conference," May 20, 1918; and "The New Map of Europe," May 8, 1919. See also "Some Geographical Aspects of International Reconstruction," *Scottish Geographical Magazine* 23 (1917): 1–11.

10. H. J. Mackinder, *Democratic Ideals and Reality: A Study in the Politics of Reconstruction*, p. 160.

gested that the book espoused "a political philosophy that appears to be out of harmony with the most hopeful tendencies of our times."[11]

Democratic Ideals and Reality consists of eight chapters. In the opening chapter, "Perspective," Mackinder stated his position in relation to what he termed the geographical realities: "Unless I wholly misread the facts of geography I would . . . say that groupings of lands and seas, and of fertility and natural pathways, is such as to lend itself to the growth of empires, and in the end of a single world-empire." In the scramble to label this statement environmental determinism, many commentators failed to notice the important riders Mackinder added. For him human victory would consist of rising above the fatalism suggested by the groupings of land and seas. "We must recognize these geographical realities and take steps to counter their influence."[12]

In the second chapter, "Social Momentum," two major themes were introduced. Drawing on the work of his colleagues J. A. R. Marriott and C. Grant Robertson, Mackinder argued that in Prussia, since the beginning of the nineteenth century, the education system had been harnessed to the needs of the state to produce a nationalistic concept of "Kultur."[13] Later in

11. *The Spectator*, September 27, 1919, p. 408; *American Historical Review* 25 (1920): 258. Earlier Teggert contributed a long, and generally favorable, review essay to a leading geography journal: "Geography as an Aid to Statecraft: An Appreciation of Mackinder's Democratic Ideals and Reality," *Geographical Review* 8 (1919): 227–42.
12. Mackinder, *Democratic Ideals and Reality*, p. 2. In his introduction to the 1962 reprint of *Democratic Ideals and Reality*, Pearce is careful to avoid labeling Mackinder a determinist.
13. Ibid., p. 19; J. A. R. Marriott and Charles Grant Robertson, *The Evolution of Prussia: The Making of an Empire*. Marriott and Robertson were well-published historians who wrote extensively on Germany. Robertson's *Bismarck* (1918) went to many editions. J. A. R. Marriott (1859–1945) became secretary of Oxford University Extension Delegacy when Michael Sadler left in 1895, and he worked closely with Mackinder in connection with Reading Extension College. Later they sat in the House of Commons together. Marriott, like Mackinder, and probably many of his generation, could remember precisely when he learned of the French surrender at Sedan in 1870 (John Marriott, *Memories of Four Score Years*, p. 18). Grant Robertson (1869–1948) was a historian with a strong interest

the book, Mackinder pointed out that the modern German states was highly centralized and controlled much economic activity, including the banks and railroads. The functions of tariffs were to protect key industries and facilitate the penetration of markets in adjoining countries. Large German corporations, often with a high degree of state involvement, had immense economic and political power at home and in neighboring lands. The whole economic and social structure—the "going concern"—had developed in order to harness human resources in service to the state. In Mackinder's view, Germany's defeat in World War I had not altered the system.[14]

Having a centralized machinery of state with the power to control citizens was bad enough, but Mackinder perceived that a new type of "organizer" was beginning to occupy leadership roles. The "organizers" were prevalent in Germany and could be seen, to a lesser degree, in Britain and France. Mackinder hinted that Joseph Chamberlain had been an example of the new type of leader with tendencies toward dictatorship.[15]

Mackinder devoted the next two chapers to "The Seaman's Point of View" and "The Landsman's Point of View." Here he set out, in historical context, the opening up of the globe as a backdrop to world affairs in the twentieth century. The Heartland was defined as:

> the regions of Arctic and Continental drainage [which] measure nearly a half of Asia and a quarter of Europe, and form a great continuous patch in the north and centre of the continent. That whole patch, extending right across from the icy, flat shore of Siberia to the torrid, steep coasts of Baluchistan and Persia, has been inaccessible to navigation from the ocean. The opening of it by railways—for it was practically roadless beforehand—and by aeroplane routes in the near future, constitutes a revolution in the relations of men to the larger geographical reali-

---

in geography. He edited historical atlases with J. G. Bartholomew. In 1919, Mackinder suggested that his name be considered as director of the School of Geography, Oxford (Mackinder to Freshfield, April 7, 1919, RGS). Robertson became principal and then vice-chancellor of Birmingham University (1920–38).

14. Mackinder, *Democratic Ideals and Reality*, pp. 142–43.
15. Ibid., p. 27, note 1.

ties of the world. Let us call this great region the Heartland of the Continent.

As compared with the Pivot of 1904, the Heartland was much enlarged. The Heartland contained areas of continental drainage excluded from the Pivot, and it included all of East Europe. World War I had shown that sea power could not penetrate the Baltic and Black seas and in practice, "The Heartland is the region to which, under modern conditions, sea-power can be refused access, though the western part of it lies without the region of Arctic and Continental drainage."[16]

The inclusion of East Europe in the Heartland concept was of importance. Mackinder, after an examination of the events leading up to World War I, had come to the opinion that the struggle for command of the Heartland would be between Germany and Russia. The crucial ground was East Europe, and for one or two generations the odds would lie with the Germans, for the Russians were "hopelessly incapable of resisting German penetration." The Heartland concept was not a restatement of the Pivot idea; it was a prediction made in the light of practical politics and the First World War, and it proved to be remarkably accurate. The Pivot statement was more theoretical and treated the problem of sea power versus land power: the whale and the bear antagonism of nineteenth-century imperialism. The 1919 statement brought in a tradition that saw Central Europe as the fulcrum from which the lever of power could be exercised. Who rules Bohemia rules Europe was how Bismarck had expressed the theme. The Masaryk articles on Pan-Germanism in *The New Europe* had made Mackinder fully aware of this line of thought in German consciousness. In 1916 Masaryk declared, "[I]f Berlin succeeds in creating 'Central Europe,' the aim of the war is attained . . . . The Great War is a daring attempt to organize Europe, Asia, and Africa—the Old World—under the leadership of Germany."[17] In *Democratic Ideals and Reality* Mackinder encapsulated this theme in his widely quoted jingle:

16. Ibid., pp. 73–74, 110.
17. Ibid., p. 158; Thomas G. Masaryk, "Pangermanism and the Eastern Question," p. 19.

Who rules East Europe commands the Heartland:
Who rules the Heartland commands the World-Island:
Who rules the World Island commands the World.[18]

By 1939 Germany had performed as Mackinder predicted and asserted control over key elements in the middle tier of states: Austria, Czechoslovakia, and Poland. In 1941, Germany invaded Russia, a move that can be interpreted as a thrust into the Heartland. Mackinder may have tended to underestimate the power of Russia in the long term, but he correctly predicted Germany's intentions in East Europe over a twenty-year period.

In chapter six, "The Freedom of Nations," Mackinder set out his view of the structure of the postwar world. It was important to settle the question between the Germans and Slavs and "see to it that East Europe, like West Europe, is divided into self-contained nations." The League of Nations should uphold international law, although there was the problem that the big members were more influential than the small states in the organization. The British Empire should evolve so that its units took their places in the League. He looked for cooperation in West Europe between Britain, France, and Italy. Mackinder hoped to see Prussia "broken into several federal states" and thought it probable "that the Russians will fall into a number of states in some sort of loose federation." As a result he wanted to see the end of the German-Russian two-state system in East Europe and the emergence of the middle-tier states supported by the League.[19]

Other suggestions included joint trusteeship over Panama, Gibraltar, Malta, Suez, Aden, and Singapore by Britain and the United States, to assure "peace of the ocean." Here Mackinder recognized the rise of the United States to great-power status in the oceanic realm. Not until 1924, in *The Nations of the Modern World*, did he mention the germ of the NATO idea, when he suggested that Western Europe and North America constituted "a single community of nations," with a common cause.[20]

18. Mackinder, *Democratic Ideals and Reality*, p. 150.
19. Ibid., pp. 157, 171.
20. Ibid., p. 172; H. J. Mackinder, *The World War and After: A Concise*

Mackinder's Suggested Boundaries for East Europe, 1919

In the first two decades of the twentieth century, Mackinder's view of the world had evolved to take account of events. He still gave more weight to Germany than Russia in any struggle for control of the World Island. He recognized the growing importance of the United States in the oceanic world. He reinforced his vision of a British Empire composed of provincial units with more local control, even reversing his previous view on Ireland and advocating independence for Southern Ireland.[21]

In his chapter six, "The Freedom of Nations," Mackinder

Narrative and Some Tentative Ideas. This point is discussed by G. R. Crone, "Sir Halford John Mackinder," *Encyclopaedia Britannica,* 1981.

21. Mackinder, *Democratic Ideals and Reality,* p. 196.

wrote with extraordinary foresight of the strategic considerations that would contribute to the outbreak of World War Two.
In chapter 7, "The Freedom of Men," he predicted the rise of
totalitarian regimes. Mackinder argued that states should consist of a combination of communities. This had been the pattern in late medieval Europe, when cities had been the centers of regions and the concept of nationality was not strongly
developed. In the nineteenth century, regional strength had
declined as a result of a number of forces. Emerging nation-
wide interests had produced horizontal cleavages in society
both nationally and internationally. International combines
of capital, class organizations, and similar institutions undermined local communities that had possessed a balanced life
of their own.[22]

Countries were distorted by centralization, which fostered
the growth of massive metropolitan areas. In England, at the
beginning of the nineteenth century, one sixteenth of the population lived in London. By the end of the First World War, one
Englishman in five lived in the capital. The development of
rail and road networks, radiating from London, resulted in the
capital "milking the country."[23] The best talent was drawn into
the metropolitan area, to the detriment of outlying regions.

The process of centralization, which still was going on, had
been hastened by a conflict that encouraged states to take additional powers to organize the war effort. Added to this, "our
Western communities are passing through a dangerous stage
in this generation. Half-educated people are in a very suscep-
tible condition and the world today consists mainly of half-
educated people. They are capable of seizing ideas, but they
have not attained the habit of testing them."[24]

When "The Freedom of Men" is considered alongside the
chapter on "Social Momentum," a frightening prospect is apparent. States had become highly centralized and were in danger of being controlled by ruthless organizers who would be
able to manipulate the belief systems of half-educated popu-

22. G. R. Crone, "Sir Halford John Mackinder," *Encyclopaedia Britan-
    nica*, 1981; Mackinder, *Democratic Ideals and Reality*, p. 268.
23. Mackinder, *Democratic Ideals and Reality*, p. 191.
24. Ibid., p. 187.

lations. Unfortunately, Mackinder did not coin a vivid phrase to represent these fears. Reviewers found it easy to pick up the Heartland concept but failed to heed the terrible warning that states were likely to fall into the control of dictators. Within twenty years, Russia, Italy, Germany, and Spain would be organized as he feared.

In talks to the London School of Economics and the Royal Glasgow Philosophical Society in 1921, Mackinder spoke in more detail of the centralization of power. Under the title "The New Methods of Long Distance Communication and Their Probable Effects," he suggested that the influence of the organizer would be greatly increased by "the power of retailing personality," which mass communication allowed. The effects of centralization were gathering pace in Britain. The four great railway companies "might be a stepping stone to a single company which the state would be compelled to dominate." Banks, newspapers, and shipping lines were controlled by a few major firms in each industry. Concentration might give economic efficiency, but if the ship owners came to run the newspapers, and the banks the shippers, then there would be "a power capable of dominating the national government for the capability would exist to bias news, suggest ideas, and control thoughts."[25]

Fears of highly centralized states controlled by ruthless organizers were not the only long-term problems suggested in *Democratic Ideals and Reality*. The thoughts on balanced local economic development were in a tradition that would culminate in the "small is beautiful" movement. Mackinder's ideas on large companies operating internationally were far ahead of general concern about the influence of multinational corporations. His notion that interest groups, working on the centralized machinery of state, might undermine democratic institutions is now a matter of concern in the United States.

Not all the blame for the neglect of *Democratic Ideals and Reality* can be ascribed to negligent reviewers and euphoric readers in the aftermath of the war to end all wars. The haste with which the book was written meant that many themes

25. *The Times*, October 7, 1921; *Glasgow Herald*, October 7, 1921, November 24, 1921.

were not fully explained. Further, because Mackinder was looking a long way into the future, he spoke of tendencies and possibilities. He never attempted to give precise dates for coming events. The book lay neglected for over twenty years, but in the 1940s, after Germany had fulfilled Mackinder's prophecy in the East, it was reprinted several times. Then the reviews were more appreciative. One reviewer even thought that the depth of understanding displayed by Mackinder allowed his work to be considered alongside that of Oswald Spengler.[26]

British High Commissioner to South Russia 1919–1920

As 1918 came to a close, Mackinder was disappointed not to be asked to take a part in the postwar boundary-making process. In 1916 he had asked Lord Milner to keep him in mind for overseas missions, pointing out that he was fluent in French and could get by in German.[27] In 1918 Milner did recommend him for the Saar Basin Delimitation Commission, but nothing came of this. In the following year, after Lord Curzon took over from Balfour as Foreign Secretary, and a few months after the publication of *Democratic Ideals and Reality* with its views on Russia's becoming a loose federation of states, a task was found for Mackinder. On October 23 he was called to the Foreign Office and asked if he would go to South Russia, as British High Commissioner, to offer assistance to General Anton Ivanovich Denikin, who was leading forces against the Bolshevik regime.[28]

26. *Atlantic* 170 (1942): 152.
27. Mackinder to Milner, December 11, 1916, Milner Papers, Letters 1916, Bodleian. Portions of this section have appeared as "Sir Halford Mackinder as British High Commissioner to South Russia, 1919–1920," *Geographical Journal* 142 (1976): 228–36.
28. The major documentary sources concerning the mission are in the Public Record Office, London (PRO); they include: CAB/24/94, Draft instructions for Mr. Mackinder, November 1919 (also published in E. L. Woodward and Rohan Butler, eds., *Documents on British Foreign Policy, 1919–39*, First Series, vol. 3, HMSO 1949); FO 800/251 Russia, Private papers of Sir H. J. Mackinder relating to his mission to South Russia October 1919–February 1920, Curzon to Mackinder, October 23, 1919; FO 800/251, Mackinder to Curzon, October 24, 1919; CAB/24/97, Report on the situation in South Rus-

During the war, after Russia and Germany signed the treaty of Brest-Litovsk (March 1918), allied troops had been stationed around Murmansk, Archangel, and Vladivostok and in South Russia to protect military supplies. From the stationing of troops it was a short step to giving tacit support to anti-Bolshevik forces, and by January 1919 the British were supplying Denikin's forces with arms. In May these forces began an offensive, which resulted in the capture of Orel, some 250 miles south of Moscow, in October 1919. At this time the British decided to send a High Commissioner and an economic advisory staff to help administer the newly won territory. Hardly had Mackinder been appointed than British policy began to swing in another direction. It was announced that no more arms would be sent to South Russia after the end of March 1920. And the capture of Orel proved to be the high tide of Denikin's advance. Soon his forces were falling back to the south.

Originally Mackinder was to leave London in November, travel overland to Brindisi, and thence by naval vessel to the Black Sea. After some delay, however, he was sent across Europe to consult with authorities in East European states, including Poland and Roumania. In Warsaw, Mackinder and the British minister, Horace Rumbold, worked out a scheme to bring the South Russian and Polish forces into a concerted effort against the Bolsheviks.[29] From Warsaw, before proceeding to Constantinople, Mackinder traveled to Bucharest and Sofia. He then sailed across the Black Sea to Novorossiysk, where he met the British representative from the Transcaucasian republics of Armenia, Azerbaijan, and Georgia. These republics aspired to independence.

On January 10, 1920, Mackinder moved inland to Tikhoretskaya Junction for meetings with General Denikin. By this time the military situation was so bad that there could be no ques-

sia, by Sir H. J. Mackinder, MP; CAB/23/20, Conclusions of a meeting of the Cabinet, held at 10 Downing Street, S.W. 1, 29 January 1920, at 11:30 A.M. More detail on sources is given in "Sir Halford Mackinder as British High Commissioner."
29. Rumbold to Mackinder, September 1, 1920, Ms Rumbold, dep. 27, Bodleian; Martin Gilbert, *Sir Horace Rumbold*, pp. 190–93.

tion of setting up a high commission. Consequently, on January 16, Mackinder began the return journey to London. He had assembled a picture of the situation from Poland to South Russia and had laid the foundation of what amounted to an anti-Bolshevik alliance. He had persuaded Denikin to recognize the *de facto* independence of border peoples and to submit the Russia-Poland border to arbitration on an ethnogeographic basis.

On his return to England, Mackinder was called to Downing Street to address the Cabinet. He described the general situation and suggested a policy that can be summarized as follows: Give all the anti-Bolshevik states, from Finland to the Caucasus, some support. Denikin should be equipped anew, for defensive purposes on a modest scale. The British must be prepared to hold the Baku-Batum rail line and to take control of Denikin's fleet on the Caspian Sea. Support for individual states was a waste of money. It was necessary to adopt the whole policy. The alternative was to see the Bolsheviks advance to the Black Sea, which would be a complete moral victory for them. It would not be necessary to drive the policy to extremes. Once an alliance had been created, and the morale of the anti-Bolshevik states reestablished, Britain should be in a better position to negotiate a peace with Soviet Russia.[30]

The implications of Mackinder's statements were farreaching. He was advocating, in addition to the new states created at Versailles, a further subdivision of East Europe and the Caucasus. By his policy a series of buffer states, perhaps White Russia, the Ukraine, South Russia, Georgia, Armenia, Azerbaijan, and Daghestan, would emerge. The danger of Russia's becoming a great Heartland power would be lessened.

By the conclusion of his mission, Mackinder realized that a Bolshevik Russia might be capable of exploiting the strategic position of the Heartland. To quote from the report to the Cabinet, Mackinder feared the rise of "a new Russian Czardom of the Proletariat" and "Bolshevism, sweeping forward like a prairie fire" towards India and what he termed lower Asia, creating a world that would be a very "unsafe place for democ-

30. CAB/23/20, PRO.

Mackinder's Suggested Boundary Changes,
East Europe and the Caucasus, 1920

racies."[31] On May 20, 1920, he made this prophecy public, in a speech to the House of Commons.

The alternatives were not simply a Bolshevik or a White government in Moscow. Mackinder recognized other possibilities. First, organized economic life might collapse in Russia, creating a condition of "well-fed barbarism." Second, if no viable central government were established in Russia, German economic and administrative expertise might fill the void. Third, the British feared an industrial resurgence in Germany, but they dared not allow that country to collapse economi-

31. CAB/24/97, PRO.

cally, thus creating the distress that might prove a fertile ground for Bolshevism.[32] The idea that Germany would hold Bolshevism in check was already emerging. As Hardinge of the Foreign Office wrote to Rumbold in the spring of 1920: "I doubt if the danger of Bolshevism spreading outwards is as great as it was at one time thought. Provided we do not upset the German government, I believe Germany will be a bulwark against Bolshevism."[33]

The policy proposals Mackinder put to the Cabinet for a system of alliances received no support. Some members were strongly opposed. H. A. L. Fisher recorded, "Cabinet on Russia — Mackinder's absurd report. As I had to leave early I let the PM know in writing that I disagree with it. Long thinks it an able effort! B. Law more sensible." Lord Curzon was absent from the Cabinet, but the Foreign Office view was represented by a deputy. Winston Churchill, when presented with a long-term policy for containing Bolshevism, did not support Mackinder's scheme. By January 1920 the Cabinet group, which earlier had supported intervention, knew it could not get policies for additional aid adopted. There had been a change of mind. As Hardinge told Rumbold:

> There had indeed been a very great change in the attitude of Europe towards the Bolsheviks, chiefly I think because it was realized that without embarking upon another war it would be impossible to defeat them. It may be short-sighted policy but in this country everybody is sick of war and the feeling is that the Bolsheviks and Russia must be left to stew in their own juice.[34]

Early in February Mackinder resigned his position as high commissioner. It had become untenable in every sense. In April 1920, Denikin resigned, and in the autumn the South Rus-

32. Stanley W. Page, *The Geopolitics of Leninism.* Engels, Marx, and Lenin all believed that Germany was likely to be the major center of communism.
33. Hardinge to Rumbold, April 9, 1920, Ms Rumbold, dep. 26, Bodleian.
34. Diary, January 29, 1920, H. A. L. Fisher Papers, Bodleian; Hardinge to Rumbold, April 9, 1920, Ms Rumbold, dep. 26, Bodleian.

sian forces were evacuated via the Black Sea. Mackinder's knighthood was announced on January 1, 1920.

Mackinder lived long enough to see many of his fears concerning Russia justified. What such statesmen as Curzon and Mackinder wished to do in 1919 was to reverse a process of expansion from the Russian core area. If an independent existence could be restored to regions that had been absorbed into Russia, increasing the size, resources, and manpower of the state, pressure could be relieved on British spheres of influence, particularly in the Indian subcontinent.

The analysis made by Mackinder was not incorrect, but the political realities of the time in a democracy made it impossible to act upon his viewpoint. Britain, tired of interventions in foreign disputes, withdrew from Russia. With her went the support that many potentially autonomous regions, such as the Ukraine, had sought.

Mackinder and Rumbold, who had laid the groundwork for the anti-Bolshevik alliance, took their reverse well, although they both felt there had been a lack of consistency and purpose in Britain. As Mackinder commented in a letter to Rumbold, he had been "up against the fact of a complete *volte face* while I was absent!"[35]

The Impact of Mackinder's Strategic Thinking

Mackinder's strategic thinking can be traced into at least three traditions: British, German, and American. His ideas were widely discussed, but to what degree they led to the emergence of policy will always be a matter for debate. In chapter seven, "The Geographical Pivot of History," it was suggested that Mackinder's ideas filtered into the Conservative party in Britain and contributed to the acceptance of the common-market concept.

In Germany, *Democratic Ideals and Reality* caused much interest in geopolitical circles. In 1925, Karl Haushofer reviewed the book in *Zeitschrift Für Geopolitic*. Haushofer saw

35. Mackinder to Rumbold, August 25, 1920, Ms Rumbold, dep. 27, Bodleian.

Mackinder's work as the "greatest of all geographical world views" and thought he had been responsible for training a generation of imperialist British statesman. Haushofer, who was highly interested in British ideas concerning the course of world events, had both Mackinder's book and James Fairgrieve's *Military Geography*, translated into German. However, Mackinder did not contribute in an original way to German ideas of geopolitik. It is true that Haushofer transmitted some knowledge to his student Rudolf Hess, but the professor was less influential than is sometimes supposed. Anyway, the whole concept of the importance of East Europe was in German strategic thinking long before the Pivot paper was published in 1904.

Mackinder was used in Germany in just the same way the American, A. T. Mahan, was used in England. He was the prestigious outsider quoted to buttress established orthodoxy. When Eyre Crowe, of the Foreign Office, wrote his famous memorandum on British sea power in 1907, he referred to Admiral Mahan in support to his view that "sea power is more potent than land power." It would be ludicrous to suggest that Crowe would not have held exactly the same opinion on sea power if Mahan had not published a word on the subject. Similarly, Mackinder's views were used to bolster long-held German opinions. There is a school of thought which believes that Mackinder was a powerful force in the development of German geopolitics. The strongest recent proponent of this view is W. H. Parker, who states, "Mackinder became one of the most influential thinkers of modern times, helping to determine the course of history through his impact upon the external policies of Germany before the war and the United States after it."[36]

In the United States, which inherited leadership of the

36. W. H. Parker, *Mackinder: Geography as an Aid to Statecraft*, p. 147. Robert B. Downs, *Books That Changed the World*, pp. 263–73, offers a similar view; in his article in the *Encyclopaedia Britannica*, Crone describes the idea that Mackinder had inspired Hitler as an "absurd notion." Barry A. Leach, *German Strategy Against Russia, 1939–41*, contains an excellent discussion of the general background to German policy in the East.

oceanic world from Britain, Mackinder's ideas have received serious study. The historian Teggart brought the Pivot paper to the notice of American academics in more than one of his books during and after the First World War.[37] *Democratic Ideals and Reality* was more widely read in the United States than in Britain, particularly by military officers. Knowledge of Mackinder's ideas among this influential group prepared the way for the postwar policy of containment (Chapter 11).

Some have suggested that Mackinder tended to overlook the strategic importance of North America. This is incorrect. Mackinder knew the United States and Canada firsthand. He traveled in the United States in 1892 and visited Canada on several occasions. He knew leading educators in both countries and in Canada had friends at many levels of society. In the 1904 statement, North America is seen as part of the oceanic realm, a realm that he predicted would decline toward the end of the Columbian epoch. North America was portrayed in a similar way in *Democratic Ideals and Reality*. After World War I, Mackinder quickly appreciated that the center of power within the oceanic realm had moved westward, and he said so in *The Nations of The Modern World* (1924). In the "Round World" statement (1943) the United States is depicted as the undoubted leader of the Western World. Even so, the passage westward of power within the oceanic realm had come about because the marginal lands were less able to resist the great land power of Eurasia. The rise of America was associated with a relative decline of the West.

In some senses the emergence of Russia as a superpower can be seen as fulfilling the ideas suggested by Mackinder in the Pivot paper. Nevertheless, Ladis Kristof, after a careful study of the relevant Russian literature, has concluded that the work of Mackinder has largely been ignored in Russia.[38] Of course,

---

37. Frederick John Teggart was born in Belfast in 1870 and educated at Trinity College, Dublin, before going to the United States in 1889. He entered Stanford in 1894 and subsequently held appointments at Berkeley. Teggart introduced Mackinder's "Geographical Pivot of History" to American historians in his books *Processes of History* (1918) and *Theory of History* (1925). He was a member of the American Geographical Society.
38. Ladis K. D. Kristof, "Mackinder's Concept of Heartland and the

the fact that ideas go unnoticed in officialdom is no indication that the processes were not operative.

In 1919, Mackinder wrote of the danger of a resurgent Germany making a thrust for world domination. In 1920 he made proposals that might have helped curtail the longer-term threat from Russia. However, what was said in *Democratic Ideals and Reality* was "out of harmony with the most hopeful tendencies of our times," and the British cabinet had no stomach for the Russian policies advocated by Mackinder in 1920. Had Mackinder's post-World War I ideas been followed, would they have changed the world? If we had been able to ask him this question in 1946, when he was 85, the idea still would have produced a vivid flash of interest in his eyes. There would have been a pause and then a measured answer delivered in slightly elliptical language. The substance might have been: "I doubt my suggestions would have changed fundamental relationships. My ideas would have prolonged the process by which the World Island power emerged, giving the oceanic world more time to respond. On the other hand, it is just possible that technological developments, emerging toward the end of the twentieth century, would have altered strategic relationships once again, developing a new situation." At the end of his life he was still capable of looking a long way forward, and he was very careful not to say, "I told you so."

Russians," paper delivered at the Twenty-Third International Geographical Congress, Leningrad, July 22–26, 1976. Parker, *Mackinder*, p. 184, suggests a different view.

CHAPTER

11
The Round World

Professor Halford Mackinder retired from the London School of Economics in 1925. The title of professor, conferred upon him in 1923, came late largely because, after his entry into Parliament in 1910, Mackinder did not take a full-time part in academic life.[1]

When Mackinder retired from London University he still had work in public life, particularly on the imperial committees and several important commissions. He became a privy councilor in 1926, the same year that the council granted a charter giving Reading full university status. Academic interests were not neglected when he left LSE. Although Mackinder did not write a major new work in the last twenty years of his life, he updated the school texts and *Britain and the British Seas*. He delivered papers at meetings and was active in the affairs of the Royal Geographical Society and the Geographical Association. By 1930, because deafness was causing him problems, Mackinder was less able to perform speaking engagements and committee assignments that took him away from familiar voices.[2]

The Imperial Committees and Retirement

The First World War had necessitated maritime cooperation between distant parts of the Empire, particularly in view of

1. E. W. Gilbert, "Halford John Mackinder," *The Dictionary of National Biography, 1941–50*, p. 556; J. F. Unstead, "H. J. Mackinder and the New Geography," *Geographical Journal* 113 (June 1949): 56–57.
2. Mackinder to A. G. Ogilvie, June 15, 1934, Sixth Floor Collection, Edinburgh University Library. Mackinder added a P.S.: [I]t is now 51 years since I got my first at Oxford on an answer—three hours long— on the geographical distribution of animals!"

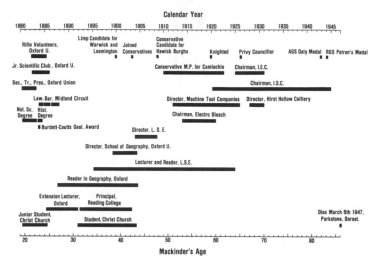

Diagram summarizing Mackinder's careers.

the German submarine threat to supplies transported to Britain. Cooperation had proved beneficial, and the participants hoped to develop additional mechanisms after the war. On July 26, 1918, the Imperial War Conference passed a resolution which proposed that "shipping on the principal routes, should be brought under review by an Inter-Imperial Board on which the United Kingdom and the British Dominions and Dependencies should be represented."[3]

On June 15, 1920, Prime Minister Lloyd George authorized the setting up of the fifteen-member Imperial Shipping Committee with Mackinder as chairman. The committee operated from the Board of Trade, whose staff provided support. The members consisted of representatives from imperial territories and men in shipping or commerce. The committee had two functions: first, to look into complaints relating to discriminatory ocean freight rates and second, to survey maritime transport on imperial routes and make recommendations for the improvement of facilities with regard to the type, size, and

3. Memorandum on origins of Imperial Shipping Committee sent to Mr. McDougall (Australia), BT/118/4, PRO.

speed of ships, the depth of water in docks and channels, and the construction of harbor works.

The committee reported to the Imperial Conference. It had no powers to enforce its view, nor had it authority to compel witnesses to attend hearings. Nevertheless, it became a permanent committee in 1923, and Mackinder was given a salary as chairman. He would serve as chairman from 1920 to 1945.

The Imperial Shipping Committee represented the type of institution needed to move toward a greater degree of economic cooperation within the Empire-Commonwealth. The committee tackled a variety of problems. It advised the authorities in East Africa, Ceylon, and Southeast Asia on whether their ports were large enough and deep enough. The hope of the local administrators was that if their harbors were deepened, more vessels would call. The committee usually advised against expensive improvements because these did not, by themselves, create additional trade. In any event, over considerable portions of the globe, the depths of the Suez and Panama canals determined the size of vessels.[4]

The committee worked on marine insurance rates, freight rates for numerous commodities, and cases in which a rate appeared to discriminate against an Empire product. For example, it cost more to ship apples to Britain from Canada than from the United States in British-owned ships. The committee tackled the question of rebates and shipping conferences and issued a report that did something to clarify the situation. There were some considerable successes. One of the first tasks the committee undertook was to look at the limitation of shipowners' liability in relation to bills of lading. In 1921 uniform legislation was recommended to governments within the Empire, and in the same year the idea was taken up by the Brussels Diplomatic Conference with a view to promoting worldwide conformity. Another useful report concerning Canadian cattle shipping did result in benefits to Canadian cattle producers and consumers in the United Kingdom.

In all, between 1920 and 1930 the committee had 150 formal meetings and numerous informal gatherings. The pub-

4. BT/188/3, PRO.

lished reports, most of which were written by Mackinder, usually contained a succinct appraisal of the problem at hand and careful advice on possible courses of action. The committee tended to favor the ship operators, who were strongly represented on the committee, when freight rates were discussed, but there is no doubt that the committee illuminated many problems.

In the 1930s the balance of the committee's functions altered. It spent less time looking at freight rates and concentrated upon its second objective of surveying maritime transport on imperial routes. At a meeting of March 20, 1934, the committee determined to undertake "a general survey of shipping within the Empire."[5]

With no thought that he was now in his seventies, Mackinder threw himself into the new project with his usual energy. The task was huge, for the survey was to be nothing less than a study of resources, trade routes, and future prospects of the shipping industry within the British Empire. Griffith Denbigh Jones, a former merchant seaman with a degree from LSE (1935) and qualifications in economics and statistics, was hired to help with the project. For every ship trading in the Empire a card was prepared giving the ship's name, flag, ownership, type, tonnage, voyages, position from time to time, and other information. By 1936, six thousand cards existed and two additional assistants were brought in.[6]

After the data-collection phase, the committee conducted hearings to determine how the shipping industry was faring. From the reports of witnesses it appeared that British shipping was doing well in most regions. However, in the Far East the Japanese were making gains at British expense. The committee read an article by G. C. Allen, "The Concentration of Economic Control in Japan," which appeared in the *Economic Journal* in June 1937. Allen had lectured on economics at the Higher Commercial College in Nagoya, Japan, from 1922 to 1925 and then had taken appointments at the Universities of Birmingham and Hull. At the University of Liverpool he had

5. BT/188/31, PRO.
6. BT/188/277, PRO.

held the Brunner Chair of Economic Science since 1933. Professor Allen was invited to London to address the committee on how large Japanese companies operated. The report on the Far East, drafted by Mackinder and the secretary of the committee, suggested that:

> Although the Japanese have learned a great deal in the way of shipbuilding and ship management from Britain they have added and made applicable, under oriental conditions of competition, what amounts almost to a new technique, based on integration, especially of merchanting with shipowning. . . . Japanese competition in the Orient is one in which the supplies of goods are massed in support of one group of shipowners in competition with other shipowners.[7]

The committee went on working until the outbreak of World War II. Although it met rarely during the war years, the information the committee had collected on imperial shipping was of value in coping with Britain's critical supply situation. Sir Halford retired from the chairmanship in 1945. The committee became the Commonwealth Shipping Committee in 1948 and did not cease to function until 1963. In a study of the Imperial Shipping Committee, Burley concluded that "much of the achievement between 1920 and 1939 was due to Mackinder's tireless energy . . . his meticulous attention to detail, and his visionary dedication to the imperial ideal. During his Chairmanship the Committee as a whole met no fewer than 233 times."[8] Out of the meetings and deliberations, thirty-nine unanimous published reports appeared. This was a considerable achievement for a committee that had little in the way of formal power. In order to obtain a unanimity of view, Mackinder was quite capable of applying pressure, and to this end he was not always entirely polite with those appearing at the meetings.

Probably arising out of his connections with the shipping industry, Mackinder was asked by Vickers Armstrong to nego-

7. G. C. Allen, "The Concentration of Economic Control in Japan," *Economic Journal* 47 (1937): 271–86. BT/188/33, PRO. BT/188/67, PRO.
8. Kevin H. Burley, "The Imperial Shipping Committee," *Journal of Imperial and Commonwealth History* 2 (1974): 212.

tiate a claim the company had against the Turkish government. Before 1914, Vickers had a concession to operate a dockyard in Turkey. Of course the concession was canceled when Britain and Turkey found themselves on opposite sides in World War I. After the war, Vickers wanted compensation for the plant, which had been taken over by the Turkish government. Arbitration was provided for in the peace treaty with Turkey, signed in 1923.

Vickers thought that its dockyard was worth half a million sterling, but by 1925, when Mackinder was brought in to negotiate for the company, the claim had been scaled down to £335,000. Mackinder negotiated with Mustafa Cheref Bey and, although the two men made some progress, matters were deadlocked in 1928. Cheref Bey was offering £60,000 on behalf of the Turkish government and, even though Mackinder had come down to less than £200,000, a settlement was no nearer, and he withdrew from the negotiations.[9]

Mackinder was active in the affairs of another imperial committee. In 1923 the Imperial Economic Conference recommended the establishment of an Imperial Economic Committee. In 1925 Sidney Webb, who was president of the Board of Trade, set up this committee with Mackinder as chairman. The role of the committee was more limited than its title suggested. The main function was to examine:

> the possibility of improving the methods of preparing for market and marketing within the United Kingdom the food products of the overseas parts of the Empire with a view to increasing the consumption of such products in the United Kingdom in preference to imports from foreign countries, and to promote the interests both of producers and consumers.[10]

The committee was highly active. Between 1925 and 1931, the end of Mackinder's term, it produced eighteen reports examining the possibility of importing more meat, fruit, fish, poultry, dairy goods, honey, tobacco, timber, pig products, rubber, hides, and tea into Britain from commonwealth countries.

---

9. Vickers Ltd., Vickers House, Millbank, London, Record 163-H.
10. CAB/24/168 C.P. 465, PRO.

Mackinder retained some business interests, but overall his income was modest. In 1939, as he approached his eightieth birthday, it was decided that he should retire from the chairmanship of the Imperial Shipping Committee and that, although he was to retain an honorary appointment, his salary would cease.[11] Now he had neither the need to live in London nor the income to sustain residence in the capital. Over the years he had found it necessary to move out from the heart of London to find cheaper accommodation. St. James Court, Westminster, had been followed by Chelsea and then Putney. Now, with war approaching and income diminishing, it was time to move again. With limited financial resources, Sir Halford had little choice but to join his retired brother, the Reverend Lionel Mackinder, and sister-in-law, Eleanor, in a modest house at Parkstone, Dorset, a small town lying beside Bournemouth.

The pace of life was slowing. There was time to reach back and touch people and institutions that had helped make his life. In 1937, at the invitation of his estranged wife, Mackinder went to Capri as part of a long visit to the Mediterranean. Bonnie lived in a small villa with her sister, Hilda Hinde, who had helped with the Kenya expedition in 1899. Halford's ten-day stay on Capri was a success. Whatever had to be said between Sir Halford and Lady Mackinder was said, and they stayed in contact for the rest of his life. As World War II approached, Bonnie, a British citizen, had to leave Italy and move to Switzerland, where Halford was able to visit her in 1939 and 1946. In the last days of his life, Halford added a codicil to his will, providing funds to repair the Capri villa, which had been damaged in the absence of Bonnie and Hilda.[12]

Late in September 1940, Mackinder went to Oxford to spend a few days with Sir Michael Sadler. The old men rejoiced in

11. When Britain entered the Second World War, Mackinder still was chairman of the ISC; his resignation did not take effect until 1945. He had made arrangements to leave London before the war started. The ISC met infrequently in the period 1939–45.
12. Mackinder to Sadler, September 29, 1940, Ms. Eng. Misc. C. 549, Bodleian; Parker, *Mackinder*, p. 55; Will of Halford John Mackinder, Probated May 20, 1947, Somerset House. The original will was drawn January 26, 1939. The codicil was added March 1, 1947.

the triumph of the Battle of Britain and, after an interval of decades, resumed an easy companionship. They talked about their successes and disappointments, including the failure of the Mackinder marriage which had disrupted Halford's life and helped drive him away from old friends. When Halford left Oxford, Sadler recorded his feelings. The visit had left a wholly pleasant impression. Sadler thought Mackinder a big man with great insight into world politics. But, Sir Michael mused, for his "ability, strength, courage he has not been placed as he should have been. Why?" Sadler obviously thought that the era in politics had not produced results commensurate with Mackinder's talents but that, whatever his past disappointments, Halford was now happy.[13]

In 1942, Mackinder visited the University of Reading and gave a talk to the students studying geography. He must have derived satisfaction at seeing the outcome of the extension college that he and Sadler had hatched, half a century before, on the street beside Keble College.

In the same year another institution he had helped create, the Geographical Association, celebrated its fiftieth year. Mackinder attended the annual conference, held at Exeter in April, and delivered a paper, "Geography, An Art and a Philosophy." Mackinder had served as president of the Geographical Association in 1916 and since that time had been chairman of the association's council. In 1917, H. J. Fleure was appointed to the honorary positions of secretary and editor to the association. Fleure and Mackinder, however, had not always enjoyed cordial relations. Fleure thought, mistakenly, that Mackinder was upset by comments he made in a review of *Nations of the Modern World*, published by Halford in 1911. "Would there were more geography and less politics!" were the sharp words, but all of this was long since forgotten as the two men worked closely to promote the work of the association.[14] Mackinder

13. Mackinder to Sadler, September 29, 1940, Mackinder to Sadler, October 3, 1940, Sadler diary, October 22, 1940, Ms. Eng. Misc. C. 549, Bodleian.
14. The substance of this address appears in H. J. Mackinder, "Geography: An Art and a Philosophy," *Geography* 27:138 (December 1942): 122–30. The review of *Nations of the Modern World* appeared in

maintained contact with the School of Geography, Oxford, via Kenneth Mason and E. W. Gilbert. Billy Gilbert, over a period of years, corresponded and conversed with Mackinder and recorded much that otherwise might have slipped away.[15]

If Halford Mackinder thought that his life was to be lived out in pleasant obscurity on the Dorset coast, punctuated only by visits to and from old friends, he experienced a surprise. *Democratic Ideals and Reality* had received little notice when published in 1919. Just one more book dealing with postwar problems, it had been ignored.[16] Events were to reveal the prophetic quality of the volume. As the new states of Eastern Europe—Czechoslovakia, Austria, and Poland—came under German control, Mackinder's ideas were rediscovered. In 1939, after Munich, the *Anschluss*, and the Nazi-Soviet pact, the *New Statesman and Nation* (August 26, 1939) carried a comment on the way Mackinder's ideas, as interpreted by the German geopolitician General Karl Haushofer, seemed to be reflected in Hitler's policies. The notion was erroneous in that Hitler's expansionist ideas had deep-seated roots in German nineteenth-century nationalistic thought. But, never mind the facts, here was the basis of a sensational story which many journalists were to work up into exciting headlines.

The verifiable parts of the story are as follows. A former professional soldier and geographer, Karl Haushofer, developed the study of what he called *geopolitik* in Munich after the First World War. Appalled at Germany's treatment at Versailles,

The Geographical Teacher 6 (1911–12): 179–80. Fleure confided to Geddes his fear that he had made an enemy of Mackinder (MS 10572 ff 68–69v, Geddes Papers, SNL). Later in life he wrote cheerfully, "I remember my irreverence in attacking Mackinder's imperialism; I think Herbertson was inclined to sympathize and Mackinder was too great to resent it" (H. J. Fleure, "Recollections of A. J. Herbertson," *Geography* 50:229 [November 1965]: 348). In a letter to T. W. Freeman (March 26, 1948), Fleure recorded his impressions of Mackinder in a forthright but friendly way.
15. Edmund William Gilbert (1900–73) was educated at Oxford, where he read history and subsequently took the diploma in geography. He taught at London and Reading before returning to Oxford in 1936. He was professor of geography from 1953 to 1967 and a fellow of Hertford College from 1939 to 1967.
16. Hilda Ormesby, *The Times*, March 17, 1947.

Haushofer began to construct arguments to set things right in his view.[17] Haushofer was not an original thinker, and he combed the work of writers like Curzon, Mackinder, Holdich, Fairgreave, Mahan, Bowman, and Kjellan (author of the term *geopolitics*) for ideas to buttress his arguments. Haushofer had several books by foreign writers translated into German, and these he reviewed in the journal *Zeitschrift für Geopolitik*, which he published in Munich from 1924.

Among Haushofer's students in Munich was Rudolf Hess, a close associate of Hitler. It has been claimed that Haushofer helped Hitler draft *Mein Kampf*. When interrogated after World War II, Haushofer flatly contradicted this, stating that the first time he saw the work was when it appeared in print. When Hitler came to power, however, Haushofer found that his work was well funded and that his ideas received wide recognition. But around 1938, Haushofer lost popularity because he did not advocate unbridled German expansion in Central and Eastern Europe. An advocate of uniting people of German language, he belonged to several *Volkdeutsch* organizations which aimed to promote the interests of Germans living beyond the boundaries of the fatherland. Between 1933 and 1938, Haushofer felt pressured into pro-expansionist statements. On November 8, 1938, he had a sharp disagreement with Hitler, arguing against further expansion.[18] He never saw Hitler again.

17. As Hartshorne has pointed out, the distinction between political geography and geopolitics is the tendency of the latter "to go beyond the study of things as they are and render judgements as to what they should be. When dealing with some problem in which his own country is concerned, the student will strive to demonstrate that things should be made as he, a loyal citizen, would like them to be" (Richard Hartshorne, "Recent Developments in Political Geography, II," *American Political Science Review* 29:6 [December 1935]: 958). In part I of the study (*American Political Science Review* 29:5 [October 1935]: 793), Hartshorne gives relatively little weight to the impact of Mackinder's ideas upon Haushofer. He thinks Fairgreave is more important. In other words, in a detached study, the foremost authority on German geography between the wars did not see Mackinder's ideas as a preponderant influence upon Haushofer. Only later, literally in the heat of battle, did the press invent the Mackinder, Haushofer, Hess, Hitler "linkage."

18. David P. Benorden, "General Karl N. Haushofer and Geopolitics," pp. 83–85; Edmund A. Walsh, *Total Power: A Footnote to History,*

The flight of Hess in 1941 robbed Haushofer of patronage, and whatever influence he held rapidly disappeared. He ended the war in Dachau, and his son Albrecht was murdered by the S.S. After 1945 it was clear that Haushofer had never held anything like the influence ascribed to him by the allied press during the war. His influence had been "more a product of American imagination than actual fact."[19]

Little of this was known in 1941, when war fever hit the United States. The idea implanted by the *New Statesman* became the basis of an industry that flourished for several years, spawning articles and books which formed a spectrum from the spurious to the scholarly. In February 1941, *Newsweek* pronounced that the whole Nazi war scheme was based upon the ideas of an Englishman, Sir Halford Mackinder. In June of the same year, *Reader's Digest* reproduced from *Current History* an article claiming that the one thousand scientists of Haushofer's Geopolitical Institute in Munich "dominate Hitler's thinking." Geopolitics became an American fad. Magazines and journals commissioned articles on the subject. In 1942, *Life* carried a long article on geopolitics that set out Mackinder's ideas and went on to make the familiar claim that Mackinder had set down a blueprint for German expansion.[20] Many academic journals carried substantial articles analyzing geopolitics, and several important books appeared, including works by R. Strausz-Hupe and Hans Weigert.[21] Among the

344–53. Walsh took a statement from Haushofer on November 2, 1945. In chapter 3 of *Generals and Geographers: The Twilight of Geopolitics*, Hans W. Weigert pointed out, early in the war, that there were important differences in the geopolitical views of Haushofer and Hitler. In particular, they held very different views about Russia, and Haushofer advocated cooperating with the Soviet Union (pp. 153–64).

19. Thomas R. Smith and Lloyd D. Black, "German Geography: War, Work and Present Status," *Geographical Review* 36:3 (July 1946): 23.

20. *Newsweek*, February 17, 1941, p. 24; *Reader's Digest*, June 1941, p. 23; *Life*, December 21, 1942, p. 108.

21. R. Strausz-Hupe, *Geopolitics: The Struggle for Space and Power;* Weigert, *Generals and Geographers.* Examples of journal articles are: Charles B. Hagan, "Geopolitics," *Journal of Politics* 4 (1942): 478–90; George Kiss, "Political Geography into Geopolitics," *Geographical Review* 32 (October 1942): 632–45; and H. F. Raup, "Geopolitics," *Education* 63:5 (January 1943): 266–72. Raup, pp. 270–71,

new books *America's Strategy in World Politics* (1942), by Nicholas Spykman, was probably the most influential.[22]

Dutch by birth, Nicholas Spykman (1893–1943) had pursued a career as an international correspondent before going to Berkeley and gaining bachelor's, master's, and doctoral degrees in the years 1921, 1922, and 1923. He then moved to Yale to teach international relations and became director of the Institute of International Studies. With a broad experience of the world, Spykman was free of much idealized American thinking about how international affairs ought to be run. In the midst of war he looked ahead to the emerging patterns of power politics and set out in *America's Strategy* what the United States had to do to retain a balance of power. In a widely quoted phrase, Spykman announced, "A Russian state from the Urals to the North Sea can be no improvement over a German state from the North Sea to the Urals." As the United States had just embraced the Soviet Union as an ally, this type of thinking was seen in some quarters as cold-blooded power politics. Nevertheless Spykman's book was widely reviewed and widely read by policymakers.[23] It rose above the flood of jour-

---

makes the point that Haushofer probably disapproved of the German attack on Russia.

22. Nicholas J. Spykman, *America's Strategy in World Politics*, received a front-page review in the *New York Times Book Review*, April 19, 1942.

23. Spykman defended his views on the balance of power in a letter to *Life*, January 11, 1943, in which he argued that "justice is most likely to prevail among states of approximately equal strength, and democracy can be safe only in a world in which the growth of unbalanced power can be effectively prevented." Alfred E. Eckes, Jr., *The United States and the Global Struggle for Minerals*, pp. 106–107, states that Spykman was "read widely in policymaking circles." Spykman did not quote Mackinder in *America's Strategy*, although he implicitly accepted the idea of a Eurasian heartland and used the term *heartland* frequently (pp. 180–86). Edward Mead Earle, "Power Politics and American World Policy," *Political Science Quarterly* 58:1 (March 1943): 102, pointed out that Spykman did not always acknowledge where he drew on the "pioneer work of others in his field." Of course, Spykman was writing in haste to get his ideas in front of a broad reading public at an opportune time. In the posthumously published *Geography of the Peace*, Spykman analyzed Mackinder's ideas and suggested that the rimland of Eurasia would be the crucial area after World War II. Mackinder's per-

nalistic output concerning geopolitics and indicated that, whether America liked it or not, there was cause for concern about the broader political alliances that would emerge after World War II.

Foreign Affairs

Mackinder's work was caught up in the debates about grand strategy and geopolitics. In 1942, at the urging of Edward Mead Earle, Henry Holt reissued *Democratic Ideals and Reality* with new front material by Earle and George Fielding Eliot. Mackinder made no revisions but contributed a short opening note.[24]

Earle had served in World War I and then devoted himself to the study of history and world affairs. He taught at Columbia before joining the newly created Institute for Advanced Study at Princeton in 1934. Several departments of government sought his views, and he was a frequent lecturer at the Army War College. By the beginning of the war he was attached to the office of Strategic Services and was a special advisor to the air force. Earle and other scholars at the Institute for Advanced Study, along with their counterparts at the Yale Institute of International Studies, formed a think tank on major strategic questions. That Mackinder's *Democratic Ideals and Reality* was reintroduced by Earle to American readers indicated that it was being carefully studied by scholars well connected with policy-making debates.

The reappearance of *Democratic Ideals and Reality* and the number of articles which mentioned Mackinder's work caught the interest of *Foreign Affairs*. Isaiah Bowman, who served on

---

sistent interest in a balance of power in "The Round World" article (1943) parallels Spykman's thoughts in *America's Strategy*. No correspondence between Spykman and Mackinder has been traced, although Spykman did reference Mackinder's work.

24. Halford J. Mackinder, *Democratic Ideals and Reality: A Study in the Politics of Reconstruction.* Mackinder's copy of the reprint is in the Geographical Association Library, Sheffield. The copy has been carefully corrected. It would make an excellent starting point for a reprint of *Democratic Ideals and Reality*, which currently is out of print.

the editorial board and had the ear of Roosevelt on geostrategic questions, suggested to the editor that he ask Mackinder
for a contribution. This idea was endorsed by Hans Weigert
who had corresponded with Mackinder earlier in the year.[25]

On December 21, 1942, Hamilton Fish Armstrong, editor
of *Foreign Affairs*, wrote to Mackinder and asked him to prepare an article which would carry further some of the themes
of *Democratic Ideals and Reality*. Delayed as a result of the
war, the letter took more than a month to arrive, but Mackinder accepted promptly and set to work on an article entitled
"The Round World and the Winning of the Peace." By February 26, 1943, a first draft had been dispatched across the Atlantic in the official mails.[26] Within a few days Mackinder sent
the first of several revisions, and Fish Armstrong made the suggested changes. Armstrong then asked Mackinder to compose
a few sentences on how he visualized future relations between
Russia and the West. To this Mackinder replied that he expected Germany to be overlooked by great powers on both sides
and that cooperation between the West and Russia would be
a safeguard to peace.[27]

"The Round World and the Winning of the Peace" was published in *Foreign Affairs*, July, 1943.[28] In many ways the article served to update Mackinder's thinking on world strategy,

25. Eckes, *The United States and the Global Struggle for Minerals*,
    p. 290, n. 32. Weigert, *Generals and Geographers*, p. 5, refers to let-
    ters from Mackinder to Weigert in January and June, 1942. I have
    not been able to trace this correspondence.
26. Armstrong to Mackinder, December 21, 1942, Mackinder to Arm-
    strong, February 26, 1943, Hamilton Fish Armstrong Papers, Prince-
    ton University Library (HFA). As a privy councillor and chairman
    of the Imperial Shipping Committee, Mackinder was able to get
    one of the secretaries of state to send the material in a diplomatic
    bag. I am indebted to William P. Bundy, editor of *Foreign Affairs*,
    for locating and copying the Armstrong-Mackinder correspondence.
27. Mackinder to Armstrong, March 1, 1943, Mackinder to Armstrong,
    received April 23, 1943, Armstrong to Mackinder, May 18, 1943,
    Armstrong to Mackinder, May 24, 1943, Mackinder to Armstrong,
    telegram, no date, HFA.
28. Halford J. Mackinder, "The Round World and the Winning of the
    Peace," *Foreign Affairs* 21 (1943): 595–605. A slightly revised ver-
    sion was reprinted in Hans W. Weigert and Vilhjalmur Stefansson,
    eds., *Compass of the World: A Symposium of Political Geography*.
    Bowman, Hamilton, and Weigert had persuaded Mackinder to write

but it contains delphic passages not easy to interpret. In sum the article does not carry much comfort for the West.

[T]he conclusion is unavoidable that if the Soviet Union emerges from the war as conqueror of Germany she must rank as the greatest land power on the globe. . . . [S]he will be the power in the strategically strongest defensive position. The Heartland is the greatest natural fortress on earth. For the first time in history it is manned by a garrison sufficient both in number and in quality.[29]

At the end of the war, Germany would be embanked between the Heartland and the amphibious powers—America, Britain, and France. Going back to an idea he proposed in 1924, Mackinder suggested that the West would organize around the North Atlantic with a bridgehead in France, a "moated aerodrome" in Britain, and a reserve of manpower, industries, and resources in the United States and Canada. This notion did find expression in NATO. However, although Hamilton pressed him on the issue, Mackinder did not offer any answer on how the Atlantic powers would relate organizationally to Russia except in their mutual desire to resist any future German threat.[30] The paper carries an assumption, however, that the Atlantic powers and the Soviet Union will remain allies. "The Round World" ends with a hint that, if the Atlantic powers and Soviet Union fall out, it may be that the Asiatic peoples will come to control the Heartland.[31]

"The Round World" is strongly anti-German, as Mackinder had been at least since the beginning of the century. The article is not anti-Russian, and the Soviet Union is seen as forming an essential part of an alliance which will maintain a balance of power and keep Germany in a peaceful role. The message seems to be cooperate with Russia in order to main-

---

down his thoughts on the origin of the Pivot idea and express views on the form of the postwar world.

29. *Foreign Affairs*, 1943, p. 605.
30. H. J. Mackinder, *The World War and After*, pp. 251–52; Armstrong to Mackinder, May 18, 1943, Armstrong to Mackinder, May 24, 1943, HFA.
31. For an important discussion of this point, see W. H. Parker, *Mackinder*, pp. 210–12.

tain a balance of power against a dimly perceived threat of Asiatic power.

Mackinder and the Policy of Containment

During the war years the American reading public, exposed to discussion of geostrategic questions, developed an awareness of the broader strategic problems in Europe and Asia. Military magazines, in particular, had excellent coverage of grand strategy and freely quoted the work of Spykman and Mackinder. The idea of a strategic Heartland with the potential to dominate the Eurasian landmass was reiterated to the point of banality between 1941 and 1944.

In his posthumous book *Geography of the Peace* (1944), Spykman laid down the broad geographical patterns that could form the basis of a containment policy. Around the Heartland, he suggested, was a Rimland stretching from Europe, through the Middle East, to South Asia and the Far East. It would be in the Rimland that the struggle for control of Eurasia would take place. Postwar American policy—NATO, CENTO, SEATO, the Korean conflict, and treaties with Japan—might be viewed as an attempt to ally with Rimland states to contain the Heartland power.

The policy of containment is regarded as having different roots. In February 1946, George Kennan, after a careful study of Russian actions, sent the famous long telegram from Moscow to Washington, in which he argued for a much firmer U.S. policy towards the Soviet Union. The telegram was followed by his influential article in *Foreign Affairs* entitled "The Sources of Soviet Conduct." This work is generally acknowledged to represent the immediate origins of the policy of containment within the State Department.[32]

Containment had a prehistory outside the foreign service. In the years 1941–44 the American public became aware of

32. *Foreign Affairs* 25 (1947): 566–82. John Lewis Gaddis, *Strategies of Containment*, pp. 3–24, points out that there were several influential figures within the foreign service, including William C. Bullitt and Averell Harriman, who wanted a firmer policy on Russia prior to Kennan.

the need for a continental policy. When Kennan's ideas circulated within government circles they necessarily coalesced with the concept of a Eurasian Heartland. Kennan had a policy to deal with a problem that was already widely perceived in defense circles—how the United States would cope with the power that controlled the Heartland. The military embraced containment eagerly, and among serving officers the geostrategic concepts of Mackinder were widely known. Mackinder did not sow the seed of containment, but his works helped prepare the ground in which the idea germinated.[33]

Summary

We arrive at the end of Mackinder's professional career knowing little of the man. Although he worked with a large number of people, he had few close friends other than Thomas Walker, Michael Sadler, and William Childs. Few people could write of the inner man from firsthand knowledge. Essentially rather shy and private, Mackinder was driven by a necessity to keep proving himself to himself. In 1916, when he was in his midfifties, an age at which many men are content with their role in the world, he told Lord Milner he still had not done the best that was in him.[34] This phrase helps us understand why Mackinder was constantly moving to a new career. He always felt the need to accept the next challenge: to start a discipline in a university, to found a university, to implement educational reform, and to attempt to restructure the Empire in order to secure the place of Britain in the world. The Oxford reader in geography became the principal of Reading, the director of LSE, the advocate of imperial unity, and then a member of Parliament and a Commonwealth statesman.

The progression was not simply ambitious careerism, for

33. Albert C. Wedermeyer, *Wedermeyer Reports*, pp. 51–53. There is no evidence that Mackinder's ideas were ever studied as a direct foundation for foreign policy. Richard Hartshorne and Kirk Stone (personal communications) have no recollection of any such work in OSS; Robert Strauz-Hupe (personal communication) feels that Mackinder's ideas played only a modest role in shaping American policy.
34. Mackinder to Milner, December 11, 1916, Milner Papers, Bodleian.

behind the whole development was a visionary element. Mackinder saw the world as a place in which the biggest landmass was liable to give rise to a large, highly centralized state, in which provinces and individuals would be submerged and controlled. Most of what Mackinder feared, namely the rise of big, highly centralized states capable of threatening the world, came to pass. His hopes for a multiethnic Commonwealth, a league of democracies capable of protecting itself, made little progress.

Although many people in the early years of the century saw parts of what Mackinder visualized, no one else, in practical politics, had quite so complete a picture of what the future held for Britain and how threats might be met. The Round Table group worked for a multiethnic British Commonwealth of Nations, and there were innumerable advocates of imperial defense, imperial unity, and tariff reform. Many perceived the problem of the rising military power of continental states. But Mackinder came closest to seeing how the world would develop politically, militarily, and economically in the first half of the twentieth century.

Mackinder suggested that the age of sea power was coming to an end, that the balance would shift in favor of continental powers. Britain would face economic problems as well. Although British banking and commerce would remain strong, manufacturing industry was in decline relative to other industrial powers. With two of Britain's major advantages, sea power and industrialization, being eroded, the future looked uncertain, particularly in view of the likely emergence of a great land power in Eurasia. What was to be done? Mackinder's answer was to create a league of democracies from the Empire. But to organize another massive state was not the object. The parts had to have provincial identity and mutual respect. Above all, the provincial populations had to be well educated and knowledgeable about each other.

It is one thing to make observations about the long-term trend of events and quite another to persuade institutions to prepare for the changes. Mackinder tended to be isolated by his vision. Highly capable men who were comfortable trying to look a few years ahead were made uneasy by his views. Mackinder paid a large price for his efforts to promote a broader understanding of Britain's long-term difficulties. The shrewd careerist would have consolidated the position at Oxford and Reading and not moved to London. After creating the School of Geography and the university college at Reading he had an excellent chance of becoming vice-chancellor of one of the large provincial universities that received charters in the first decade of the century. His name was mentioned at Birmingham in this context.[35] Instead, he moved into commonwealth education, with extensive work for the Colonial Office visual-instruction committee, and became active in the imperial unity political group. One result of constantly moving to a new activity was that he never accumulated much money. This aspect of life was not important to him. From all his careers, books, and business ventures, he retained little, and when he died he left only a few thousand pounds.

As a leader, in his early years, Mackinder was exhilarating to work with. He had the ideas, the vision, and the driving power to get things done. He could delegate, had trust in colleagues, and gave them an important place in what was going on. William Childs summed it all up when he wrote that Mackinder was a

talker, convincing and provocative. He had a way of blending dreams and hard sense, subtlety and simplicity, and he never seemed to know when he passed from the one to the other. He made some opponents, as a leader in stark earnest is bound to do. He sometimes ploughed ahead, leaving a wake of troubled waters, and he certainly gave the rest of us plenty to think and talk about. Masterful, he yet made us his partners. We could always speak our minds; our criticisms were considered; sometimes they were even acted upon.[36]

Mackinder was considerate of colleagues and worked hard to see that they got security and advanced their careers. When he decided someone had a contribution worth making, he was

35. Sonenschein to Sadler, February 12, 1899, Ms. Eng. Misc. C. 551, Bodleian.
36. W. M. Childs, *Making a University*, p. 11.

patient, even if there were disappointments along the way. The best-known example of this patience is his nursing of Childs at Reading through his first year or two of mediocre performances. Childs was so appreciative of this that he could tell the story without rancor. Sometimes faith was misplaced. Dickson had not been an easy colleague at Reading but, because he was making advances in meteorology, Mackinder invited him to join the School of Geography. In a small group he did considerable damage and was partly responsible for Mackinder's being forced to make the choice between Oxford and the London School of Economics. Even this did not leave Mackinder embittered. After Herbertson's death the Royal Geographical Society asked Mackinder's opinion on several candidates who might be considered for the vacant readership. Dickson's name was in contention and Mackinder gave a fair, well-proportioned account of his abilities. There was no attempt to take revenge.

In public life Mackinder received a number of honors. He was elected to Parliament, knighted, and became a privy councilor in 1926. Honors in academia were few. London University gave him the title of professor in 1923, but there was no honorary fellowship at his Oxford College, no honorary degree from Reading University, and he never became president of the Royal Geographical Society. The Patron's medal, which the Society did award in 1945, came late, after memories apparently had been jogged in 1944, when the American Geographical Society awarded the Daly medal to Mackinder. The contrast between public honor and academic neglect was a result of the nature of the man. His achievements were substantial, but his isolation, after the early years of the century, meant there were few close friendships and few disciples of the type who were in a position to make nominations for honorary degrees or high office in professional societies.

When he looked back on life, Mackinder saw 1899, when he climbed Mount Kenya, as his culminating year. By then he had started the School of Geography, Oxford, and established the foundations of a university at Reading. He was associated with the London School of Economics, *Britain and the British Seas* was part written, and the Pivot paper was in his mind. Within a short time he was separated from his wife, Reading, and the School of Geography. Choosing to direct LSE rather than the School of Geography cost him his studentship at Christ Church and access to senior common room life. Away from Oxford he lost something. When, in later life, he jotted passages for an autobiography he wrote fondly of the camaraderie of the University Museum, the Union, friendships in Sadler's group, and the support he received in the Christ Church senior common room, as he worked to found Reading University. The move to London eventually placed him more in politics than academia. The political world, with its alliances rather than friendships, did not provide the interaction with small groups of individuals out of which most of his best ideas had grown. As a member of Moseley's first class, Mackinder recalled "the joy of comradeship in a beginning." He may have been shy, but in his years at Oxford an infectious enthusiasm overcame this, and he did work well with a great range of people. "Mackinder had a genius for taking help where he could find it," was the view of the ancient historian, John L. Myres, who had a long association with the School of Geography. The staffing of Reading was brilliantly done, given that it was an institution without prestige and money.[37]

The London School of Economics was largely a night-school operation with a part-time staff. There was little in the way of corporate life, and Mackinder's social life lacked the variety he had enjoyed in Oxford. His shyness became more pronounced, and he became in some ways a different man. There are dangers in taking the opinions of one or two people from afar, but in 1933 Miss Mactaggart of LSE looked back, and recorded her impressions of Mackinder:

> He had an enormous grip on things. He never had justice done to him: he did not leave any kind of popularity behind him. This was perhaps due to the fact that he was a very shy man: he used to dash upstairs . . . and never look at anybody. His home circumstances were unhappy, and that may have something to do

37. M.P. Auto.; J. L. Myres to T. W. Freeman, September 10, 1950. I am indebted to Walter Freeman for making this letter available.

with it. . . . . He was not very punctual. He never came in when he said he would. One day Lord Milner, who had an appointment with him, had to spend half an hour waiting for him. And once, when he did not come to give a lecture (he frequently did that) Mr. Reeves [Director after Mackinder] wanted to "sack" him on the spot and it had to be pointed out to him that Mr. Mackinder had been appointed by the University. . . . . He appreciated the first rate members of his staff. He was not jealous: he wanted people to give their best.[38]

The thought of Mackinder scuttling upstairs to his office, not wanting to look people in the eye, is in marked contrast to most accounts of the man. Those who recorded memories of him speak consistently of his erect bearing, flashing eyes, and strong presence. There seems, then, to have been a phase in the early years of the century when Mackinder was in a very difficult patch. Health problems necessitated hospitalization in 1907 and extended rest in 1909.

Once Mackinder was elected to Parliament and started the Electro Bleach company, his love of beginnings revitalized him. In the years just before World War I the students and junior staff at the London School of Economics remembered him for his wonderful lecturing ability and powerful presence in the lecture hall. They admired him, but few were close enough to know him.

Little is known of the Mackinder marriage, but it is difficult not to think that its failure played a large part in Halford's problems in the early years of the century. Bonnie Mackinder's health was uncertain from the first years of the marriage. She was prone to suffer stress and in the end found a secluded life in Capri more manageable than living in Oxford married to a man who spent a great part of his time in Reading or London. Separation came in 1900. The failure, like any failure, haunted Mackinder, and he never fully relinquished responsibility for his wife's welfare.

Was it just ill luck that two very bright people married each other when they might have done better with more staid part-

38. Recollections of Miss Mactaggart, February 11, 1933, *Material on the History of the School*, R (S.R.) 1101, BLPES.

ners who would have provided secure domestic lives? Perhaps, but the Mackinder line was becoming extinct. Halford's father, Draper, had six offspring. By the year of his ninetieth birthday, 1908, only Halford, Lionel, and Violet were alive. Neither of the sons was survived by progeny.

Whatever the reasons, after 1905, Mackinder did not achieve the type of success he had been able to produce in earlier years. In part, of course, no one could hope to go on at the earlier pace, and the provincials, like Mackinder and Sadler, found it much more difficult to influence events at the heart of the system in London, where traditional interest groups were dominant. In a sense Sadler was fortunate. He lost in politics early, went off to a professorship at Manchester, and became vicechancellor at Leeds and then master of University College, Oxford. Mackinder had some success in politics, but he had no power base and was working for a cause that eventually failed. As a result, much of Mackinder's great talent was poorly invested in the middle decades of his life. He was not naturally equipped to be a politician. He was equipped to be an original thinker. Politics took more than its fair share of his talents; as a result, *Democratic Ideals and Reality* is a highly stimulating book but not the great one he might have written.

It is, of course, ridiculous to suggest that a man who played such an important part in founding an academic discipline and a university was a failure. However, the cause he devoted the greatest attention to—the balanced development of a British Commonwealth of Nations which would retain a major place in world affairs—was unsuccessful. One reviewer of *Democratic Ideals and Reality* summed up the book in terms that could serve as an epitaph for Sir Halford's political career: "Mr. Mackinder is right, profoundly right, but he will not get many supporters."[39]

How near to understanding events Mackinder had been was revealed in the years 1939–47. In 1939 the Soviet Union and Germany, preparatory to their battle for East Europe and the Heartland, signed a nonaggression pact: a truce rather than a treaty. In September 1939, Germany invaded Poland; Russia,

39. *Saturday Review,* April 19, 1919.

under the terms of the pact, occupied the eastern part of the country. In June 1941 the battle for the Heartland started when Germany invaded Russia. The attack was launched when Germany had conquered, or allied herself to, the newly created states of East Europe. Thus, according to the Heartland thesis of 1919, Germany held the advantage. Looked at from the purer perspective of the 1904 paper, Russia held the Pivot and thus the advantage—and the Pivot did prove to be a more effective advantage than East Europe. Events were aided by the British, and their allies, following the traditional nineteenth century policy of siding with the second strongest power, in this case Russia. Once the war was over the greatest part of Germany was quickly brought back into the Western fold to help counterbalance the new strongest power—Russia.

By 1945 Mackinder's prediction of the rise of massive land power in Eurasia had been fullfilled, even if his 1919 variant scenario, which suggested Germany would be that power, had narrowly failed. Mackinder left no last testament, but certainly he derived no satisfaction from being proved right. If he looked back to the years before World War I, to the time when he saw so clearly a pattern of events involving the rise of a great Eurasian power, and the eclipse of Britain as a major force in the world, he probably felt a jab that had been familiar to him throughout life. The jab activated his fear of failure. The sense was more acute because failure was a fact and not just a fear. Mackinder had foreseen the future of Britain economically and militarily. He had entered politics to influence events, and success had been limited.

The last months of Mackinder's life reflected these personal and national failures. Early in 1947 Britain fell into the grip of one of the coldest winters of the century, as Siberian air dominated the country. Blizzardlike conditions penetrated far into the south of England. Even the Dorset coast, around Mackinder's retirement home, was covered with snow and ice. The severe weather placed a huge strain on Britain's economy, there were power cuts, and industrial production was disrupted. Britain had become a second-class power, unable to support Greece against communist rebels. In response to this withdrawal the

United States proclaimed the Truman Doctrine as part of the containment policy. The cold war had begun.

Mackinder reached the age of eighty-six on February 15, but he had become prone to chest infections, and when he contracted another, in the harsh winter, he could not shake it off. He died on March 6, 1947.

Bibliography

Works by H. J. Mackinder

1877

"A Glimpse of A.D. 1950." *The Epsomian* 7 (February).

1880

"Geological Epsom." *The Epsomian* 10 (January and March).

1885

Syllabus of a Course of Eight Lectures Delivered at Bath by Mr. H. J. Mackinder, B.A., Late Junior Student of Christ Church, on "Wealth and Wages." Oxford: Oxford University Extension.

Syllabus of a Course of Lectures on the World: The Battlefield of Wind, Water and Rocks. Oxford: Oxford University Extension.

1886

The New Geography. Oxford: Oxford University Extension.

1887

"On The Scope and Methods of Geography." *Cambridge Review* 8:247–49, 264–67.

"On The Scope and Methods of Geography." *Proceedings of the Royal Geographical Society,* New Series 9:141–60. Discussion of February 14, 1887, is reported on pp. 160–74.

"The Teaching of Geography at the Universities." *Proceedings of the Royal Geographical Society,* New Series 9:698–701.

1888

"Geographical Education: The Year's Progress at Oxford." *Proceedings of the Royal Geographical Society,* New Series 10:531–33.

"Note on Geographical Terminology." *Proceedings of the Royal Geographical Society,* New Series 10:732–33.

Syllabus of a Course of Lectures on the History and Geography of International Politics. Oxford: Oxford University Extension.
Syllabus of a Course of Lectures on Physiography. Oxford: Oxford University Extension.
Syllabus of Home Reading in Geography. Oxford: Alden.

1889

"Four Lectures on the Teaching of Geography: A Summary." *Education Times and the Journal of the College of Preceptors* 52: 504.
"Geographical Education: The Year's Progress at Oxford." *Proceedings of the Royal Geographical Society,* New Series 11:502–503.

1890

"Geographical Education: The Year's Progress at Oxford." *Proceedings of the Royal Geographical Society,* New Series 12:419–21.
"On the Necessity of Thorough Teaching in General Geography As a Preliminary to the Teaching of Commercial Geography." *Journal of the Manchester Geographical Society* 6:1–6.
"The Physical Basis of Political Geography." *Scottish Geographical Magazine* 6:78–84.
"Remarks at the Presentation of the Training College Prizes." *Proceedings of the Royal Geographical Society,* New Series 12: 476–77.
With M. E. Sadler. *University Extension: Has It a Future?* London: Cassell.

1891

"Geographical Education: The Year's Progress at Oxford." Proceedings of *The Royal Geographical Society,* New Series 13:428–29.
With M. E. Sadler. *University Extension: Past, Present and Future.* London: Cassell.

1892

"The Education of Citizens." *University Extension Journal* 1:245–49.
"Geographical Education: The Year's Progress at Oxford." *Proceedings of the Royal Geographical Society,* New Series 14:398–400.

Syllabus of a Course of Six Lectures on Revolutions in Commerce. Philadelphia: American Society for the Extension of University Teaching.

1893

"Educational Lectures." *Geographical Journal* 1:157–58.
"Reports on Geography at the Universities: Oxford." *Geographical Journal* 2:25–27.

1894

"Geography at the Universities: Oxford." *Geographical Journal* 4:29–30.
Review of Reclus' "Universal Geography." *Geographical Journal* 4:158–60.

1895

"Address to the Geography Section of the British Association." *Scottish Geographical Magazine* 11:497–511.
"The Case for a Treasury Grant Restated." *University Extension Journal* 4:6–7.
Discussion of Educational Papers, *Report Proceedings Sixth International Geographical Congress* London: Murray.
"A French Educational Congress." *University Extension Journal* 4:24–25.
"Geography at the Universities: Oxford." *Geographical Journal* 6:25–36.
"Modern Geography, German and English." Presidential Address to Section E (Geography). *Report of the Sixty-Fifth Meeting of the British Association for the Advancement of Science.* London: John Murray.
"Modern Geography, German and English." *Geographical Journal* 6:367–79.
"The Relation Between University Extension Teaching and Secondary Education." *Bryce Commission Report on Secondary Education* 5:302–304.

1896

Discussion of H. R. Mill, "Proposed Geographical Description of the British Isles Based on the Ordnance Survey." *Geographical Journal* 7:357–58.

Discussion of J. E. Marr, "The Waterways of the English Lakeland." *Geographical Journal* 7:624–25.

1897

"Geography at the Universities: Oxford." *Geographical Journal* 9:653–54.

1898

"Report on Geography at Oxford." *Geographical Journal* 12:8–9.

1899

"Geography at the Universities: Oxford." *Geographical Journal* 14:87–88.

1900

"The Ascent of Mt. Kenya." *Alpine Journal* 20:102–10.

"The Great Trade Routes." *Journal of the Institute of Bankers.* The four lectures in the series were published separately. The epitome of the first lecture appeared in 21 (January): 1–6; lecture 2 appeared in 21 (March): 137–46; lecture 3, in 21 (March): 147–55; lecture 4, in (May): 266–73.

"A Journey to the Summit of Mt. Kenya, British East Africa." *Geographical Journal* 15:453–86.

Stanford's New Orographical Maps. Compiled under the Direction of H. J. Mackinder: *The British Isles; Europe.* London: Edward Stanford.

1902

Britain and the British Seas. London: Heinemann; New York: D. Appleton.

1903

"Geography in Education." *Geographical Teacher* 2:95–101. (Reprinted in *Journal of Geography* 2:499–506.)

"Higher Education." *The Nation's Need.* Edited by S. Wilkinson. London: Constable.

1904

"The Development of Geographical Teaching out of Nature Study." *Geographical Teacher* 2:191–97.

"The Geographical Pivot of History." *Geographical Journal* 23: 421–37. (Reprinted in H. J. Mackinder, *The Scope and Meth-*

ods of Geography and the Geographical Pivot of History. London: Royal Geographical Society, 1969.)

1905

"Man-Power as a Measure of National and Imperial Strength." *National Review* 45:136–43. *Seven Lectures on the United Kingdom.* London: George Philip.
Stanford's New Orographical Maps. Compiled under the Direction of H. J. Mackinder. *Africa.* London: Edward Stanford.

1906

Money-Power and Man-Power: The Underlying Principles Rather Than the Statistics of Tariff Reform. London: Simkin-Marshall.
Our Own Islands; An Elementary Study in Geography. London: George Philip. (14th ed., 1921.)
Stanford's New Orographical Maps. Compiled under the Direction of H. J. Mackinder. *Palestine; Asia; North America.* London: Edward Stanford.

1907

Address Delivered on 10 January, 1907, on the Occasion of the Opening of the Class for the Administrative Training of Army Officers. London: HMSO.
Britain and the British Seas, 2d ed. Oxford: Clarendon Press.
"On Thinking Imperially." In *Lectures on Empire.* Edited by M. E. Sadler. London: Privately printed.
Stanford's New Orographical Maps. Compiled under the Direction of H. J. Mackinder. *South America.* London: Edward Stanford.

1908

"The Advancement of Geographical Science by Local Scientific Societies." *Naturalist* 614:70–74.
"The Geographical Environment of Great Britain." *Encyclopedia Americana.*
Lands beyond the Channel: An Elementary Study in Geography. London: George Philip.
The Rhine: Its Valley and History. London: Chatto and Windus; New York: Dodd, Mead.
Stanford's New Orographical Maps. Compiled under the Direction of H. J. Mackinder. *Australasia.* London: Edward Stanford.

1909

Discussion of W. M. Davis, "The Systematic Description of Land Forms." *Geographical Journal* 34:320–21.

"Geographical Conditions Affecting the British Empire, I: The British Islands." *Geographical Journal* 33:462–76.

"The Geographical Conditions of the Defence of the United Kingdom." *National Defence* 3:89–107.

1910

Distant Lands: An Elementary Study in Geography. London: George Philip.

India. Eight Lectures Prepared for the Visual Instruction Committee of the Colonial Office. London: George Philip.

Introduction to E. Smith, *The Reigate Sheet of the One-Inch O.S.: A Study in the Geography of the Surrey Hills.* London: Black.

1915

"Andrew John Herbertson." *Geographical Teacher* 8:143–44.

"The New Map." *Glasgow Herald*, January 30.

1916

Constitutional Problems: Speeches at a Conference Held between Representatives of the Home and Dominion Parliaments, 28 July and 2 August, 1915. London: Empire and Parliamentary Association.

"Presidential Address to the Geographical Association 1916." *Geographical Teacher* 8:271–77.

1917

"Adriatic Question." *Glasgow Herald*, December 3.

"Discussion of the Resolutions of the Five Associations." *Geographical Teacher* 9:46–53.

"Some Geographical Aspects of International Reconstruction." *Scottish Geographical Magazine* 33:1–11.

"This Unprecedented War." *Glasgow Herald*, August 4.

1918

"End of an Empire: The Break-up of Austria-Hungary." *Glasgow Herald*, October 31.

"The New Map of Europe." *Glasgow Herald*, May 8.

"Rome Conference." *Glasgow Herald,* May 20.

"The Taxation of Capital." *Glasgow Herald,* January 29.

1919

Democratic Ideals and Reality: A Study in the Politics of Recon-struction. London: Constable; New York: Holt. (Reprinted, with Foreword by Anthony J. Pearce, New York: Norton, 1962.)

1921

"Geography as a Pivotal Subject in Education." *Geographical Journal* 57:376–84.

"L'envoi." *Scottish Geographical Magazine* 37:77–79.

"Railways." *Glasgow Herald,* May 30.

1922

"The Sub-Continent of India." In *The Cambridge History of India.* Edited by E. J. Rapson. Vol. I. *Ancient India.* Cambridge: The University Press.

1923

"Empire Trade." *Glasgow Herald,* October 1.

1924

The Nations of the Modern World: An Elementary Study in Geography and History, After 1914. London: George Philip.

The World War and After: A Concise Narrative and Some Tentative Ideas. London: George Philip. (The text is the same as in *Nations of the Modern World.*)

1925

"The English Tradition and the Empire: Some Thoughts on Lord Milner's Credo and the Imperial Committees." *United Empire* 14:1–8.

1927

Discussion of J. W. Gregory, "The Fiords of the Hebrides." *Geographical Journal* 69:214–15.

1930

"The Content of Philosophical Geography." Presidential address to Section D (Human Geography). *Report of the Proceedings of the International Geographical Congress* (Cambridge, 1928). Cambridge: The University Press.

"Mount Kenya in 1899." *Geographical Journal* 76:529–34.

"Recent Economic Developments in the Dominions, Colonial and Mandated Territories." *Journal of the Royal United Services Inst.* 75:254–65.

1931

Discussion of S. W. Wooldridge and D. J. Smetham, "The Glacial Drifts of Essex and Hertfordshire, and Their Bearing upon the Agricultural and Historical Geography of the Region." *Geographical Journal* 78:268–69.

"The Human Habitat." *Records of the British Association for the Advancement of Science.* London. (Also published in *Scottish Geographical Magazine* 47:321–35.)

1932

Speech at the Anniversary Dinner of the RGS, in *Geographical Journal* 80:189–92.

1933

"The Empire Marketing Board: The Attitude of the Dominions." *United Empire* 24:508–509.

1935

"The Crown." *United Empire* 26:502.

Foreword to N. Mikhaylov, *Soviet Geography.* London: Methuen.

"Progress of Geography in the Field and in the Study during the Reign of His Majesty King George the Fifth." *Geographical Journal* 86:1–12.

1937

"The Music of the Spheres." *Proceedings of the Royal Philosophical Society, Glasgow* 63:170–81.

1942

Democratic Ideals and Reality. (Reprint with an introduction by Edward Meade Earle, New York: Holt.)

"Geography, An Art and a Philosophy." *Geography* 27:122–30.

1943

"Global Geography." *Geography* 28:69–71.

"The Round World and the Winning of the Peace." *Foreign Affairs* 21:595–605.

1944

Democratic Ideals and Reality. (Reprint, Harmondsworth: Penguin Books.)

Speech at the Presentation of the Medals at the American Embassy by the American Ambassador. *Geographical Journal* 103:132–33.

1945

Speech on Receiving the Patron's Medal of the RGS. *Geographical Journal* 105:230–32.

Other Works Cited

Allen, G. C. "The Concentration of Economic Control in Japan." *Economic Journal* 47 (1937): 271–86.

Amery, Julian. *Joseph Chamberlain and the Tariff Reform Campaign.* Vol. 5 of *Life of Joseph Chamberlain.* London: Macmillan, 1969.

Amery, L. S. *My Political Life,* vol. 1: *England before the Storm, 1896–1914.* London: Hutchinson, 1953.

Ashby, Eric, and Mary Anderson. *Portrait of Haldane.* London: Macmillan, 1974.

Barnes, J., and D. Nicholson. *The Leo Amery Diaries,* vol. 1: *1896–1929.* London: Hutchinson, 1980.

Beckwith, Ian. *History of Transport and Travel in Gainsborough.* Gainsborough: U.D.C., 1971.

Benorden, David P. "General Karl N. Haushofer and Geopolitics." M.A. thesis, University of Nebraska, 1983.

Beveridge, W. *The London School of Economics and Its Problems, 1919–37.* London: Allen and Unwin, 1960.

———. *Power and Influence: An Autobiography.* London: Hodder and Stoughton, 1953.

Blouet, B. W. "H. G. Wells and the Evolution of Some Geographic Concepts." *Area* 9 (1977): 49–52.

———. "Halford Mackinder's Heartland Thesis: Formative Influences." *Great Plains–Rocky Mountain Geographical Journal* 5 (1976): 2–6.

———. "The Maritime Origins of Mackinder's Heartland Thesis." *Great Plains–Rocky Mountain Geographical Journal* 2 (1973): 6–11.

————. *Sir Halford Mackinder, 1861–1947: Some New Perspectives*. Research Paper 13, Oxford School of Geography. Oxford, 1975.

————. "Sir Halford Mackinder as British High Commissioner to South Russia, 1919–1920." *Geographical Journal* 113 (1977): 228–36.

Bowen, E. G. "The Geography of Nations." *Geography* 48 (1963): 1–17.

Burley, Kevin H. "Canada and the Imperial Shipping Committee." *Journal of Imperial and Commonwealth History* 3 (1975): 349–63.

————. "The Imperial Shipping Committee." *Journal of Imperial and Commonwealth History* 2 (1974): 206–25.

Busch, Briton Cooper. *Britain and the Persian Gulf, 1894–1914*. Berkeley: University of California Press, 1967.

Caine, Sir Sidney. *The History of the Foundation of the London School of Economics and Political Science*. London: London School of Economics, 1963.

Calder, Kenneth J. *Britain and the Origins of New Europe, 1914–1918*. Cambridge: The University Press, 1976.

Cantor, L. M. "Halford Mackinder: His Contribution to Geography and Education." M.A. thesis, University of London, 1960.

————. "The Royal Geographical Society and the Projected London Institute of Geography, 1892–1899." *Geographical Journal* 128 (1962): 30–35.

Carter, C. C., and C. McGregor, "Long Vacation Course, Oxford School of Geography." *Geographical Teacher* 2 (1902): 172–79.

Chamberlain, A. *Politics from Inside: An Epistolary Chronicle*. New Haven: Yale University Press, 1937.

Childs, Hubert. *W. M. Childs: An Account of His Life and Work*. Reading: H. Childs, 1976.

Childs, W. M. *Making a University: An Account of the University Movement at Reading*. London: Dent, 1933.

Cohen, S. B. *Geography and Politics in a World Divided*. New York: Oxford University Press, 1973.

Cook, D. P. *Benjamin Kidd: Portrait of a Social Darwinist*. Cambridge: The University Press, 1984.

Crone, G. R. "A German View of Geopolitics." *Geographical Journal* 111 (1948): 104–108.

————. "Mackinder, Sir Halford John." *Encyclopaedia Britannica,* 15th ed.

Curzon, George N. "Central Asian Railway in Relation to the Commerical Rivalry of England and Russia." *Report of the Fifty-Ninth Meeting of the British Association for the Advancement of Science.* London: John Murray, 1890.

————. *Frontiers.* Oxford: Oxford University Press, 1907.

————. *Persia and the Anglo-Persian Question.* London: Longmans, 1892.

————. *Problems of the Far East.* London: Longmans, 1894.

————. *Russia in Central Asia in 1889 and the Anglo-Russian Question.* London: Longmans, 1892.

Darwin, Charles. *The Origin of Species.* New York: Mentor Books, 1958. (Originally published in 1859.)

Denikin, A. *The White Army.* Translated by C. Zvegintzov. London: Cape, 1930.

Dilks, David. *Curzon in India.* Vol. 1. New York: Taplinger, 1969.

Downs, Robert B. *Books That Changed the World.* Chicago: American Library Association, 1978.

Dryer, C. R. "Mackinder's 'World Island' and Its American 'Satellite.'" *Geographical Review* 9 (1920): 205–207.

Earle, E. M. *Makers of Modern Strategy: Military Thought from Machiavelli to Hitler.* Princeton, N.J.: Princeton University Press, 1943.

East, W. G., and A. E. Moodie. "How Strong Is the Heartland?" *Foreign Affairs* 29 (1950): 78–93.

Eckes, A. E. *The United States and the Global Struggle for Minerals.* Austin, Texas: University of Texas Press, 1979.

English, J. S. *Halford J. Mackinder (1861–1947).* Gainsborough: Gainsborough Public Library, 1974.

Fairgrieve, J. *Geography and World Power.* London: University of London Press, 1915.

Fest, W. *Peace or Partition: The Habsburg Monarchy and British Policy 1914–1918.* (New York: St. Martin, 1978).

Firth, C. H. *The Oxford School of Geography.* Oxford: Blackwell, 1918.

Fisher, C. A. "The Britain of the East." *Modern Asian Studies* 2 (1968): 343–76.

————. "The Changing Significance of the Commonwealth in

the Political Geography of Great Britain." *Geography* 48 (1963): 113–29.

Fischer, Fritz. *Germany's Aims in the First World War.* New York: Norton, 1967.

Fleure, H. J. "Recollections of A. J. Herbertson." *Geography* 50 (1965): 348–49.

————. "Sixty Years of Geography and Education." *Geography* 38 (1965): 231–64.

Ford, P., and G. Ford. *A Breviate of Parliamentary Papers, 1900–1916.* Oxford: Blackwell, 1957.

————. *Breviate of Parliamentary Papers, 1917–1939,* Oxford: Blackwell, 1951.

Froude, J. A. "England and Her Colonies." *Fraser's Magazine,* New Series 1 (1870): 16.

Freeman, T. W. *A History of Modern British Geography.* London: Longmans, 1980.

————. "The Royal Geographical Society and the Development of Geography." *Geography Yesterday and Tomorrow.* Edited by E. H. Brown. Oxford: Oxford University Press, 1980.

Gaddis, John Lewis. *Strategies of Containment.* New York: Oxford University Press, 1982.

Gamble, Andrew. *Britain in Decline.* Boston: Beacon Press, 1981.

Garnett, A. "Some Climatological Problems in Urban Geography with Reference to Air Pollution." *Institute of British Geographers, Transactions* 42 (1967): 21–43.

Geikie, Archibald. *An Elementary Geography of the British Isles.* London: Macmillan, 1888.

George, H. B. *Relations of History and Geography.* Oxford: Clarendon Press, 1901.

Gilbert, E. W. *British Pioneers in Geography.* Newton Abbott: David and Charles, 1972.

————. "Geography at Oxford and Cambridge," *Oxford Magazine,* February 14, 1957.

————. "The Right Honourable Sir Halford J. Mackinder, P.C., 1861–1947," *Geographical Journal* 110 (1947): 94–99.

————. "The Right Honourable Sir Halford J. Mackinder, P.C., 1861–1947," *Geographical Journal* 127 (1961): 27–29.

————. *"The Scope and Methods of Geography" and "The Geographical Pivot of History" by Sir Halford Mackinder. Reprinted with an Introduction by E. W. Gilbert*, Royal Geographical Society, London, 1951.

————. "Sir Halford John Mackinder." *Dictionary of National Biography, 1941–50.* Oxford: Oxford University Press, 1975.

————. *Sir Halford Mackinder 1861–1947: An Appreciation of His Life and Work.* London: Bell, 1961.

Gilbert, E. W., and W. H. Parker. "Mackinder's 'Democratic Ideals and Reality' after Fifty Years." *Geographical Journal* 135 (1969): 228–31.

Gilbert, Martin. *Sir Horace Rumbold.* London: Heinemann, 1973.

Goldsmid, F. J. Presidential Address, Section E, *Report of the Fifty-Sixth Meeting of the British Association Held at Birmingham, September, 1886.* London: John Murray, pp. 721–26.

Gollwitzer, Heinz. *Europe in the Age of Imperialism, 1880–1914.* London: Thames and Hudson, 1969.

Gooch, G. P., and Harold Temperley, eds. *British Documents on the Origins of the War, 1898–1914.* Vol. 3. London: HMSO.

Goudie, A. S. "George Nathaniel Curzon: Superior Geographer." *Geographical Journal* 146 (1980): 203–209.

Graham, G. S. *Great Britain in the Indian Ocean, 1810–1850.* Oxford: Clarendon Press: 1967.

————. *The Politics of Naval Supremacy.* Cambridge: The University Press, 1965.

Hall, A. H. "Mackinder and the Course of Events." *Annals of the Association of American Geographers* 45 (1955): 109–26.

Hartshorne, Richard. "Recent Developments in Political Geography, II." *American Political Science Review* 29 (1935): 785–804.

————. "Recent Developments in Political Geography, II." *American Political Science Review* 29 (1935): 943–66.

Hayek, F. A. "The London School of Economics, 1895–1945." *Economica.* New Series 13 (1946): 1–31.

Henrikson, A. K. "America's Changing Place in the World: From 'Periphery' to 'Centre.'" *Spatial Variations in Politics.* Edited by J. Gottmann. London: Sage, 1980.

Herbertson, A. J. "The Major Natural Regions: An Essay in Systematic Geography." *Geographical Journal* 25 (1905): 300–12.

Hewins, W. A. S. *Apologia of an Imperialist: Forty Years of Empire Policy.* 2 Vols. London: Constable, 1929.

————. *The London School of Economics and Political Science.* Oxford: Privately printed, 1895.

Heyes, J. F. "A Plea for Geography." *The Oxford Magazine.* December 8, 1886.

Hinde, H. B. *The Masai Language.* Cambridge: The University Press, 1901.

————. *Vocabularies of the Kamba and Kikuyu Languages of East Africa.* Cambridge: The University Press, 1904.

Hinde, Sidney Langford. *The Fall of the Congo Arabs.* London: Methuen, 1897.

————. "Three Years' Travel in the Congo Free State." *Geographical Journal* 5 (1895): 426–46.

Hinde, S. L., and Hildegarde Hinde. *The Last of the Masai.* London: Heinemann, 1901.

Holt, J. C. *The University of Reading: The First Forty Years.* Reading: Reading University Press, 1977.

Holt-Jensen, Arild. *Geography: Its History and Concepts.* Totowa, N.J.: Barnes and Noble, 1980.

Hooson, D. J. M. "A New Soviet Heartland?" *Geographical Journal* 128 (1962): 19–29.

————. *A New Soviet Heartland?* Princeton, N.J.: Van Nostrand, 1964.

House, J. W. *The Geographer in a Turbulent Age.* Oxford: Clarendon Press, 1976.

James, Preston E., and Geoffrey J. Martin. *All Possible Worlds: A History of Geographical Ideas.* New York: John Wiley, 1981.

Kadish, A. *The Oxford Economists in the Late Nineteenth Century.* New York: Oxford University Press, 1983.

Kearns, Gerry. "Halford John Mackinder." *Geographers Biobibliographical Studies,* vol. 9. Edited by T. W. Freeman. London and New York: Mansell Publishing, 1985.

Keltie, J. S. *Geographical Education.* Royal Geographical Society, Supplementary Papers. Vol. 1. 1886.

Kendle, J. E. *The Colonial and Imperial Conferences, 1887–1911: A Study in Imperial Organization.* London: Longmans, 1967.

Kennan, G. "The Sources of Soviet Conduct." *Foreign Affairs* 25 (1947): 566–82.

Kennear, Michael. *The Fall of Lloyd George.* London: Macmillan, 1973.

Kennedy, P. M. *The Rise and Fall of British Naval Mastery.* London: Allen Lane, 1976.

———. *Strategy and Diplomacy, 1870–1945.* London: Fontana, 1984.

Kristof, L. K. D. "The Origins and Evolution of Geopolitics." *Journal of Conflict Resolution* 4 (1960): 15–51.

Leach, Barry A. *German Strategy Against Russia, 1939–41.* Oxford: Clarendon Press, 1973.

Livezey, W. E. *Mahan on Sea Power.* Norman: University of Oklahoma Press, 1947.

Lorne, Marquis of. *Imperial Federation.* London: Sonnenschein, 1885.

———. Presidential Address, *Proceedings of the Royal Geographical Society* 8 (1886): 421–22.

Lyde, L. W. *Some Frontiers of Tomorrow: An Aspiration for Europe.* London: Black, 1915.

———. "Types of Political Frontiers in Europe." *Geographical Journal* 45 (1915): 126–45.

Lyttleton, E. *Alfred Lyttleton: An Account of His Life.* London: Longmans, Green, 1917.

Mackenzie, Norman. *The Letters of Sidney and Beatrice Webb.* 3 vols. Cambridge: The University Press, 1978.

Mackenzie, Norman, and Jeanne Mackenzie. *The Fabians.* New York: Simon and Schuster, 1977.

Mackinder, Draper. *My Recreation.* Gainsborough: Privately printed, 1908.

———. "Report for September, October, November 1857, Gainsborough." *Sanitary Review and Journal of Public Health* 3 (1857): 413–15.

McLean, Ian. *The Legend of Red Clydeside.* Edinburgh: John Donald, 1983.

Macmillan, Harold. *At the End of the Day, 1961–1963.* London: Macmillan, 1973.

McNamara, Robert S. "The Military Role of Nuclear Weapons." *Foreign Affairs* 62 (1983): 59–80.

Magnus, Philip. *Kitchener: Portrait of an Imperialist.* London: Penguin, 1982.

Malin, James C. "The Contriving Brain as the Pivot of History. Sea, Landmass and Air Power: Some Bearings of Cultural Technology upon the Geography of International Relations." In *Issues and Conflicts.* Edited by G. L. Anderson. Lawrence, Kansas: University of Kansas Press, 1959.

———. *History and Ecology.* Lincoln: University of Nebraska Press, 1984.

———. "Space and History: Reflections on the Closed-Space Doctrines of Turner and Mackinder and the Challenge of Those Ideas by the Air Age." *Agricultural History* 18 (1944): 65–74.

Marriott, J. A. R., and Charles Grant Robertson. *The Evolution of Prussia: The Making of an Empire.* Oxford: Clarendon Press, 1915.

Marriott, John. *Memories of Four Score Years.* London: Blakie, 1946.

Martin, G. J. "Political Geography and Geopolitics." *Journal of Geography* 58 (1959): 441–44.

Masaryk, T. G. "The Literature of Pangermanism." *The New Europe* 1 (1916): 57–60, 89–92, 118–24, 152–57, 247–29.

———. *The New Europe.* London: Wyre and Spottiswoode, 1918.

———. "Pangermanism and the Eastern Question." *The New Europe* 1 (1916): 2–19.

Matthew, H. C. G. *The Liberal Imperialists: The Ideas and Politics of a Post-Gladstonian Elite.* London: Oxford University Press, 1973.

Meinig, D. W. "Heartland and Rimland in Eurasian History." *Western Political Quarterly* 9 (1956): 553–69.

Meyer, H. C. "Mitteleuropa in German Political Geography." *Annals of the Association of American Geographers* 36 (1946): 178–94.

Mikhaylov, N. *Soviet Geography: The New Industrial and Economic Distributions of the U.S.S.R.* With a foreword by the Rt. Hon. Sir Halford J. Mackinder. London: Methuen, 1935.

Mill, H. R. Discussion of H. J. Mackinder, "Progress of Geography during the Reign of His Majesty King George the Fifth." *Geographical Journal* 86 (1935): 13.

———. *Hugh Robert Mill: An Autobiography.* London: Longmans, 1951.

———. *The Record of the Royal Geographical Society.* London: Royal Geographical Society, 1930.

————. Review of H. J. Mackinder, "Britain and the British Seas." *Geographical Journal* 19 (1902): 489–95.

————. "The Vertical Relief of the Globe." *Scottish Geographical Magazine* 6 (1890): 182–87.

Mills, D. R. "The U.S.S.R.: A Re-Appraisal of Mackinder's Heartland Concept." *Scottish Geographical Magazine* 72 (1956): 144–53.

Mitchell, Peter Chalmers. *My Fill of Days.* London: Faber, 1937.

Moodie, A. E. *Geography behind Politics.* London: Hutchinson, 1949.

Moore, Reverend C. *History of Gainsborough.* Gainsborough: Caldicot, 1904.

Morgan, K. O. *Consensus and Disunity: The Lloyd George Coalition Government, 1918–1920.* Oxford: Clarendon Press, 1979.

Moseley, Henry Nottidge. *Notes by a Naturalist: An Account of Observations Made During the Voyage of H.M.S. "Challenger" Round the World in the Years 1872–1876.* London: Murray, 1892.

————. *Oregon: Its Resources, Climate, People, and Productions.* London: Stanford, 1878.

Moss, R. P. "Authority and Charisma: Criteria of Validity in Geographical Method." *South African Geographical Journal* 52 (1970): 13–37.

Murray, John. "Inaugural Address to the 1942 Conference of Geographical Association." *Geography* 27 (1942): 117–21.

Namier, Julia. *Lewis Namier: A Biography.* London: Oxford University Press, 1971.

Nairn, Ian, and Nikolaus Pevsner. *Surrey.* 2d ed. Harmondsworth: Penguin, 1971.

Naumann, Friedrich. *Central Europe.* New York: Knopf, 1917.

Oman, Sir Charles. *Memories of Victorian Oxford.* London: Methuen, 1941.

Ormsby, H. "The Rt. Hon. Sir Halford J. Mackinder, P.C., 1861–1947," *Geography* 32 (1947): 136–37.

Page, Stanley W. *The Geopolitics of Leninism.* New York: Columbia University Press, 1982.

Parker, Geoffrey. *Western Geopolitical Thought in the Twentieth Century.* New York: St. Martin's, 1985.

Parker, W. H. *Mackinder: Geography as an Aid to Statecraft.* Oxford: Clarendon Press, 1982.

———. *Superpowers: The United States and the Soviet Union Compared.* London: Macmillan, 1972.

Partsch, J. *Central Europe.* London: Heinemann; New York: Appleton, 1903.

Pearce, A. J. Introduction to H. J. Mackinder, *Democratic Ideals and Reality.* New York: Norton, 1962.

Pevsner, Nikolaus. *Lincolnshire.* Harmondsworth: Penguin Books, 1964.

Porter, Dilwyn. "The Unionist Tariff Reformer, 1903–1914." University of Manchester, Ph.D. dissertation, 1976.

Radice, Lisanne. *Beatrice and Sydney Webb: Fabian Socialists.* New York: St. Martin's Press, 1984.

Raverat, Gwen. *Period Piece: A Cambridge Childhood.* London: Faber and Faber, 1952.

Rempel, Richard A. *Unionists Divided: Arthur Balfour, Joseph Chamberlain and the Unionist Free Traders.* Hamden, Conn.: Archon Books, 1972.

Ronaldshay, Earl J. *The Life of Lord Curzon.* Vol. 2. London: Benn, 1928.

Russell, B. *Autobiography, 1872–1914.* Boston: Little Brown, 1967.

Salmon, Michael A. *Epsom College, 1855–1980.* Oxford: Privately printed, 1980.

Scally, Robert. *The Origins of the Lloyd George Coalition: The Politics of Social Imperialism, 1900–1918.* Princeton, N.J.: Princeton University Press, 1975.

Scargill, D. I. "The RGS and the Foundations of Geography at Oxford." *Geographical Journal* 142 (1976): 438–61.

Searle, G. R. *The Quest for National Efficiency.* Berkeley: University of California Press, 1971.

Seeley, J. R. *The Expansion of England.* London: Macmillan; Boston: Roberts, 1883.

Semmell, B. *Imperialism and Social Reform.* New York: Anchor Books, 1968.

———. "Sir Halford Mackinder: Theorist of Imperialism." *Canadian Journal of Economic and Political Science* 24 (1958): 554–61.

Senger, Robert, II. *Alfred Thayer Mahan*. Annapolis, Maryland: Naval Institute Press, 1977.

Seton-Watson, Hugh, and Christopher Seton-Watson. *The Making of a New Europe: R. W. Seton-Watson and the Last Years of Austria-Hungary*. Seattle: University of Washington Press, 1981.

Seton-Watson, R. W. *Europe in the Melting Pot*. London: Macmillan, 1919.

Sharpe, R. Bowdler, H. J. Mackinder, Ernest Saunders, and C. Camburn. "On the Birds Collected During the Mackinder Expedition to Mount Kenya." *Proceedings of the Zoological Society*, pt. 3 (1900): 596–609.

Smith, Thomas R., and Lloyd D. Black. "German Geography: War Work and Present Status." *Geographical Review* 36 (1946): 398–408.

Spykman, N. J. *America's Strategy in World Politics*. New York: Harcourt Brace, 1942.

———. *The Geography of the Peace*. New York: Harcourt Brace, 1944.

Stamp, L. D. "Obituary of H. J. Mackinder." *Nature* 154 (1947): 530–31.

Stenton, Doris H. "Frank Merry Stenton, 1880–1967." *Proceedings of the British Academy* 54 (1968): 315–423.

Stoddart, D. R. "The RGS and the Foundation of Geography at Cambridge." *Geographical Journal* 141 (1978): 216–39.

———. "The RGS and the 'New Geography': Changing Aims and Changing Roles in Nineteenth Century Science." *Geographical Journal* 146 (1980): 190–202.

———. "The Victorian Science: Huxley's 'Physiography' and Its Impact on Geography." *Institute of British Geographers, Transactions* 66 (1975): 17–40.

Strausz-Hupe, R. *Geopolitics: The Struggle for Space and Power*. New York: Putnam, 1942.

Sykes, Alan. *Tariff Reform in British Politics, 1903–1913*, Oxford: Clarendon Press, 1979.

Taylor, A. E. "William George de Burgh, 1866–1943." *Proceedings of the British Academy* 29 (1943): 371–91.

Teggart, F. J. "Geography as an Aid to Statecraft: An Appreciation of Mackinder's *Democratic Ideals and Reality." Geographical Review* 8 (1919): 227–42.

―――. *Processes of History.* New Haven, Conn.: Yale University Press, 1918.

―――. Review of *Democratic Ideals and Reality. American Historical Review* 25 (1920): 258–59.

―――. *Theory of History.* New Haven, Conn.: Yale University Press, 1925.

Thomas, Oldfield. "List of Mammals Obtained by Mr. H. J. Mackinder During His Recent Expedition to Mount Kenya, British East Africa." *Proceedings of the Zoological Society,* pt. 1 (1900): 173–80.

Topley, W. *The Geology of the Weald.* London: Geological Survey, 1875.

Troll, C. "Geographic Science in Germany During the Period 1933–45: A Critique and Justification." *Annals of the Association of American Geographers* 39 (1949): 99–137.

Ullman, R. H. *Britain and the Russian Civil War.* Princeton, N.J.: Princeton University Press, 1968.

Unstead, J. F. "H. J. Mackinder and the New Geography." *Geographical Journal* 113 (1949): 47–57.

Van Der Poel, Jean, ed. *Selections from the Smuts Papers.* Vol. 6. Cambridge: The University Press, 1973.

Walsh, Edmund A. *Total Power: A Footnote to Published History,* Garden City, N.Y.: Doubleday, 1948.

Wantage, Harriet Sarah. *Lord Wantage, V.C., K.C.B.: A Memoir by His Wife.* London: Smith Elder, 1907.

Warrington, T. C. "The Beginnings of the Geographical Association." *Geography* 38 (1953): 221–30.

Webb, B. *Diaries, 1912–1924.* Edited by M. I. Cole. 2 vols. London: Longmans, Green, 1952.

―――. *Our Partnership.* Edited by B. Drake and M. I. Cole. London: Longmans, Green, 1948.

Weigert, H. W. *Generals and Geographers: The Twilight of Geopolitics.* New York: Oxford University Press, 1942.

Wells, H. G. *Anticipations of the Reactions of Mechanical and Scientific Progress upon Human Life and Thought.* London: Chapman and Hall, 1902.

———. *Experiment in Autobiography.* 2 vols. London: Gollancz, 1934.

———. *The New Machiavelli.* New York: Duffield, 1921.

Wedermeyer, Albert C. *Wedermeyer Reports.* New York: Henry Holt, 1958.

Whebell, C. F. J. "Mackinder's Heartland Theory in Practice Today." *Geographical Magazine* 42 (1970): 630–36.

Wilkinson, Henry Spenser. *Thirty-Five Years.* London: Constable, 1933.

Woodward, E. L., and R. Butler, eds. *Documents on British Foreign Policy, 1919–1939,* First Series. Vol. 3. London: HMSO, 1949.

Wooldridge, S. W., and Frederick Goldring. *The Weald.* London: Collins, 1953.

Index

Halford Mackinder was composed into type on a Compu-graphic phototypesetter in nine and one-half point Trump with two and one-half points of spacing between the lines. Trump Gravure was selected for display. The book was designed by Cameron Poulter, composed by Metricomp, Inc., printed by Thomson-Shore, Inc., and bound by John H. Dekker & Sons. The paper on which this book is printed bears acid-free char-acteristics, for an effective life of at least three hundred years.

TEXAS A&M UNIVERSITY PRESS : COLLEGE STATION